BOTANICAL RESEARCH AND PRACTICES

AUXIN

PRODUCTION, BIOSYNTHESIS AND ROLE IN PLANT DEVELOPMENT

BOTANICAL RESEARCH AND PRACTICES

Additional books in this series can be found on Nova's website under the Series tab.

Additional e-books in this series can be found on Nova's website under the e-book tab.

BOTANICAL RESEARCH AND PRACTICES

AUXIN

PRODUCTION, BIOSYNTHESIS AND ROLE IN PLANT DEVELOPMENT

ROSARIO D. PACOVSKY
EDITOR

nova publishers
New York

Library of Congress Cataloging-in-Publication Data

Auxin : production, biosynthesis and role in plant development / editor: Rosario D. Pacovsky.
pages cm. -- (Botanical research and practices)
ISBN 978-1-61122-060-5 (hardcover)
1. Auxin. I. Pacovsky, Rosario D. II. Series: Botanical research and practices.
QK898.A8A884 2014
571.7'42--dc23
2014029138

Published by Nova Science Publishers, Inc. † New York

CONTENTS

PREFACE

The word auxin has a Greek origin: auxein means to enlarge or to grow. Auxins are an important category of plant hormones, regulating many aspects of plant growth and development throughout the plant cell cycle, from cell division, cell elongation and vascular differentiation to root initiation, somatic embryogenesis, apical dominance, tropistic responses, flowering, fruit ripening, senescence, parthenocarpy, dormacy and pollen maturation. Common purification procedures for the determination of auxins include column chromatography, solid phase extraction, liquid-liquid and mass spectrometric detection. This book discusses the production, biosynthesis and role of auxin in plant development. The topics include the effects of inoculation of auxin producing bacteria on root development in model plants and semihardwood olive cuttings; auxin's role in plant defense; auxin's influence on plant developmental responses during salinity; the history, synthesis, transport, determination, metabolism, properties and physiological effects of auxins in plant growth; and the inclusion compounds of auxins in cyclodextrins.

Chapter 1 – Organic cultivation has nowadays spread out around the world owing to its environment preservation, contrary to the overuse of pesticides and fertilisers. Plant growth-promoting bacteria (PGPB) produce indole-3 acetic acid (IAA), a plant hormone implied in root development. In this study, several bacterial isolates were selected by their *in vitro* plant growth promoting features and evaluated for their effects on rooting of model plants. Both supernatants and cells were tested independently on *Vigna unguiculata* cuttings. While low concentrations of IAA produced stimulating effects on the root elongation, a higher number of short adventitious roots were displayed by cuttings inoculated with highly IAA producers. The selected strains inoculated on semi-hardwood cuttings of four olive cultivars (arbequina, hojiblanca, picual and koroneiki) showed different rooting abilities, and innocuous effects on plant after one year. The successful effectiveness of olive tree propagation under real nursery conditions shown by the selected PGPB, together with efficiency, ecology and plant harmlessness in a long-term confirmed its suitability in organic farming system and broaden the field of application of this method into low rooting ability cultivars and even in direct competition with the conventional propagation method.

Chapter 2 – Auxin is a well-known phytohormone that plays important roles in several physiological processes in plants. Currently, many studies have indicated that multiple hormonal signals, including auxins, are implicated in plant defense signaling pathways, but the roles of these signals in plant defense remain unclear. These phytohormones regulate the gene expression of plant defense and eventually trigger the production of defense molecules, such as phenylpropanoids, phytoalexins and pathogenicity-like proteins (PR). Studies of defense signaling against root-knot nematode invasion in resistant soybean genotype suggest that auxin and gibberellin interactions modulate ROS levels and include the participation of DELLA-like proteins, thereby possibly controlling plant immune and stress responses. Auxins promote the degradation of a family of transcriptional repressors known as Auxin/Indol-3-acetic acid (Auxin/IAA). These repressors bind to auxin repressor factors (ARFs) and inhibit the transcriptional-specific auxin response genes. Most of the total auxin in plants is found in the conjugated form and the formation of auxin conjugates is an important regulatory mechanism in the activation and inactivation of IAA. To regulate plant growth and development, auxin induces the expression of three groups of genes: Aux/IAA, *GH3* and small auxin-up RNA (SAUR) families. Auxin responsive *GH3* genes have been shown to play important roles in plant defense response in *Arabidopsis thaliana* and rice. Recently, it was shown that GH3.5 acts as a bifunctional modulator in both SA (salicylic acid) and auxin signaling during pathogen infection. *A. thaliana* auxin mutants evaluated against *Pseudomonas syringae* had compromised systemic immunity, demonstrating that indole-derived compounds, but not auxin itself, are involved in the maintenance of the systemic immunity. Furthermore, it is suggested that the repression of auxin signaling through either mutations in the auxin pathway or pharmacological interference with the auxin response impairs resistance to the necrotrophic fungi *Plectosphaerella cucumerina* and *Botrytis cinerea*. This chapter describes the influence of auxin on plant responses to stress and defense reactions, including the establishment of effective immunity in the plant-pathogen interaction and the interplay of auxin and other plant hormones.

Chapter 3 – Plants integrate intrinsic developmental programs and quickly adjust growth to environmental stimuli. Plant growth and development largely depend on the phytohormone auxin which exerts its function throughout partially redundant F-box receptors, TIR1-AFBs. Several studies using pharmacological and genetic approaches demonstrated that auxin response is down-regulated under salinity. However, how auxin regulation impacts on growth and developmental programs is still under study. Developmental plasticity of roots is modulated by auxin and also is critical for tolerance to stress. Reasonably, plant acclimation to salinity could involve root growth reprogramming depending on the intensity of the stress whereby plant growth is redirected to diminish stress exposure. It is hypothesized that stress-induced morphological responses (SIMR) are mediated by auxin-mediated redox metabolism. However, one emerging question is how salinity transduces stress signals to regulate root plasticity during plant acclimation. The entwined mechanisms involving miRNAs and Salt Overly Sensitive (SOS) pathway in the regulation of auxin homeostasis will be also discussed.

Chapter 4 – The word auxin has a Greek origin: *auxein* means to enlarge or to grow. Auxins are an important category of plant hormones, regulating many aspects of plant growth and development throughout the plant cell cycle, from cell division, cell elongation and vascular differentiation to root initiation, somatic embryogenesis, apical dominance, tropistic responses, flowering, fruit ripening, senescence, parthenocarpy, dormacy and pollen maturation. Common purification procedures for the determination of auxins include column chromatography, solid phase extraction, liquid-liquid and mass spectrometric detection. The four naturally occurring (endogenous) auxins are indole-3-acetic acid (IAA), 4-chloroindole-3-acetic acid, phenylacetic acid and indole-3-butyric acid (IBA); only these four were found to be synthesized by plants. Synthetic auxin analogs include 1-naphthaleneacetic acid (α-NAA), 2,4-dichlorophenoxyacetic acid (2,4-D) and many others. The most important member of the auxin family is IAA. Most of the IAA produced within plants is conjugated to other compounds to form esters, amides or glycosyl esters. The most commonly occurring IAA-conjugates are indole-3-acetylaspartic acid (IAAsp) and a range of IAA glucose esters (IAAGlu). Conjugation seems to be a mechanism for storing IAA in cells, stabilising the level of free auxin in the plant, and metabolising its excess. Auxin in conjugated molecules is protected from oxidative breakdown and may be released again through the action of enzymes. Natural and synthetic auxins are low-molecular weight, organic substances, containing either an indole or an aromatic ring. They are crystalline and only slightly soluble in water, but readily soluble in organic solvents (ethanol, methanol, acetone, diethyl ether and dimethyl sulphoxide, DMSO) or, as weak acids, in alkaline watery solutions. In intact plants, the highest level of IAA occurs predominantly in or near the apex and IAA is transported basipetally; thus, IAA concentration decreases from the top to the bottom. The above is especially true for monocotyledons, whereas in dicotyledons the highest concentration of IAA occurs in the subapical zone, which also grows most rapidly. Auxin is also abundant in young leaves, floral organs and developing fruits and seeds. IAA is photo-oxidized rapidly (50 % in 24 h) while IBA is oxidized slowly (10 %) and NAA is very stable. IAA concentrations are high in young, fast growing organs and decline with age. The indole amino acid, tryptophan (Trp) is the precursor of IAA. There are several possibilities as to how plants may convert Trp into IAA: via indole-3-pyruvic acid, via indole-3-acetaldoxime and *via* tryptamine. Auxins exhibit active transport in a polar manner in addition to long-distance movement via vascular tissues.

Chapter 5 – The low aqueous solubility of the auxins causes major problems in their administration. In order to overcome these problems water soluble salts of auxins are used or auxins are dissolved in ethyl alcohol (or another organic solvent). It has been shown that ethyl alcohol may have modifying effects in auxin action even at low concentration (Allerup, 1962). In addition, the alcohol evaporates during use, making the solution more and more concentrated and thus stunted or abnormal growth is observed.

An alternative way to increase the water solubility of the auxins is via their encapsulation inside suitable amphiphilic molecular capsules. Cyclodextrins (CDs) are oligosaccharides consisting of six, seven or eight α-(1→4)-linked D-glucopyranose units called α-, β- or γ-CD respectively, having a toroidal shape with a lipophilic inner cavity and a hydrophilic outer surface. As a result of their molecular structure and shape, they possess a unique ability to act as molecular containers by entrapping guest molecules in their hydrophophic cavity and thus exerting a profound effect on the physicochemical properties of guest molecules (Del Valle, 2004). Moreover, due to the weak non-covalent interactions between the host and the guest molecules, these supramolecular complexes are not fixed or permanent but rather in a dynamic equilibrium with their free components in solution, thus offering a slow release of the guests and sustaining their action.

CDs are used in agriculture, food, pharmaceuticals, cosmetics, environment protection and it is generally acceptable that the inclusion complex formation of auxins with CDs could improve their bioavailability, enhance their stability and reduce considerably the side effects following administration. The low cost, biocompatible and effective degradation makes β-CDs the best candidates for such molecular encapsulants. Commercial available derivatives of β-CDs are also very interesting because of their increased aqueous solubility and inclusion capacity.

In this chapter, studies on the molecular encapsulation of eight auxins (1-NAA, 2-NAA, 4-CPA, MCPA, 2-NOA, IBA, 2,4-D and 2,4,5-T) have been extensively reviewed. The molecules that have been used as hosts are mainly native β-CDs, methylated or hydroxypropyl β-CD derivatives and in a few cases α- or γ-CDs. The combined use of different analytical techniques has provided a complete characterization of the inclusion complexes. The structures of some of these complexes have been accurately determined in crystalline state by X-ray crystallography. NMR, spectrofluorometric and phase solubility studies have given solid evidences about the formation of the complexes in solution and estimations of the binding constants. DSC and FT-IR studies have checked the formation of a true solid inclusion compound and they have determined the stoichiometry of the formed complexes in the solid state.

This inclusion complexation could be regarded as an important step in the design of novel formulations of auxins with enhanced chemical properties and the gathered results can be regarded as an essential step to the development of controlled release agricultural formulations containing auxins. In addition, the inclusion of the auxins inside the CDs' hydrophobic cavity can be regarded as a simple prototype for ligand-enzyme systems, involving mainly van der Waals and hydrophobic interactions and therefore their structural study can provide some insight about the auxins binding into the pockets of the protein receptors.

In: Auxin
Editor: Rosario D. Pacovsky
ISBN: 978-1-61122-060-5

Chapter 1

EFFECTS OF INOCULATION OF AUXIN-PRODUCING BACTERIA ON ROOT DEVELOPMENT IN MODEL PLANTS AND SEMI-HARDWOOD OLIVE CUTTINGS

M. C. Montero-Calasanz[1,2]*, C. Santamaría[1], A. Daza[1], and M. Camacho[1]
[1]IFAPA, Centro Las Torres-Tomejil. Carretera Sevilla-Cazalla, Alcalá del Río. Spain
[2]Leibniz Institute DSMZ – German Collection of Microorganisms and Cell Cultures, Braunschweig, Germany

ABSTRACT

Organic cultivation has nowadays spread out around the world owing to its environment preservation, contrary to the overuse of pesticides and fertilisers. Plant growth-promoting bacteria (PGPB) produce indole-3 acetic acid (IAA), a plant hormone implied in root development. In this study, several bacterial isolates were selected by their *in vitro* plant growth promoting features and evaluated for their effects on rooting of model plants. Both supernatants and cells were tested independently on *Vigna unguiculata* cuttings. While low concentrations of IAA produced stimulating effects on the root elongation, a higher number of short

* Corresponding author: maria.montero-calasanz@dsmz.de.

adventitious roots were displayed by cuttings inoculated with highly IAA producers. The selected strains inoculated on semi-hardwood cuttings of four olive cultivars (arbequina, hojiblanca, picual and koroneiki) showed different rooting abilities, and innocuous effects on plant after one year. The successful effectiveness of olive tree propagation under real nursery conditions shown by the selected PGPB, together with efficiency, ecology and plant harmlessness in a long-term confirmed its suitability in organic farming system and broaden the field of application of this method into low rooting ability cultivars and even in direct competition with the conventional propagation method.

Keywords: PGPB, Vigna, Canola, IAA, auxin, organic agriculture

1. Introduction

Olea europapea L. commonly named olive tree is a crop native to the Mediterranean eco-geographical region and its products, olive oil and table olives, are key in the healthy Mediterranean diet pyramid. Its cultivation has spread out and nowadays it is farmed in places such as North and South America, southern Africa, Australia, Japan and China, reaching a surface of over 9.5 million hectares and a world production of almost 3.5 million of tonne of olive oil and over 2.5 million of tonne of table olives (International Olive Oil Council, season 2011/12). Organic agriculture continue to be the fastest growing agricultural sector around the world (USDA, 2013) and the olive grove is one of best exponents, since the conversion to organic farming is relatively simple (Parra López and Calatrava Requena, 2002).

The term plant growth promoting bacteria (PGPB) comprises those bacteria that are able to help to the plant growth by means of synthesis of antibiotics that limit damage by plant pathogens, biological activities as iron sequestration, phosphate solubilisation and production of phytohormones as indole-3-acetic acid (IAA) that stimulates root elongation (Patten and Glick, 2002; Spaepen et al., 2007; Glick, 2010), and decreasing of the growth-inhibiting levels of ethylene in plant tissues by production of 1-aminocyclopropane-1-carboxylate (ACC) deaminase (Glick et al., 1998; Penrose et al., 2001; Toklikishvili et al., 2010). The use of PGPB is an emerging technology designed to improve agricultural productivity (Abbas-Zadeh et al., 2010) and sustainability (Ramos-Solano et al., 2010).

In previous studies, we proposed an organic method to obtain olive plantlets (Montero-Calasanz et al., 2013a) based on the use of PGPB as substitute of the hormone IBA (indole-3-butyric acid), a synthetic hormone not authorised in organic agriculture (for additional details see Commission Regulations 834/2007 and 889/2008). Here, we optimised the method of selection of effective PGPB strains and studied in depth plant response to different microbial IAA concentrations. In addition, the bacterial inoculation on olive tree cuttings was improved and extended into olive cultivars with low-medium rooting ability and its long-term effects were tested, confirming the effectiveness and plant harmlessness of this alternative method in the organic propagation of olive tree.

2. Material and Methods

2.1. Bacterial Strains and Culture Conditions

Strains CT42 (DSM 25369), CT276 and CT363 (DSM 25371) were isolated from rhizosphere soil samples collected in an organic olive grove located in Mairena del Aljarafe (Sevilla). Strain M16 was obtained from the mycosphere of associated *Lactarius deliciosus* to *Pinus pinea* (Barriuso et al., 2005). *Pantoea* sp. AG9, *Chryseobacterium hispalense* AG13^T (DSM 25574; Montero-Calasanz et al. 2013b), *Chryseobacterium oleae* CT348^T (DSM 25575; Montero-Calasanz et al. 2014), *Pseudomonas koreensis* CT364 (DSM 25356) and *Azospirillum brasilense* Cd (ATCC29729) were previously reported as efficient PGPB on the rooting of semi-hardwood olive tree cuttings by Montero-Calasanz et al. (2013a). All bacterial strains were grown in Nutrient Broth (NB) medium, except *A. brasilense* Cd that was grown in modified NFb medium (Döbereiner, 1980). When needed, culture media were supplemented with 100 mg l^{-1} L-tryptophan (precursor of IAA). Bacterial cultures were incubated at 28 °C, shaking at 180 rpm for 48 h, reaching about 10^9 colony-forming unit (CFU) ml^{-1}. IAA production was assessed in the bacterial cultures prior to use in assays.

2.2. In Vitro Analysis of Putative Plant Growth Promoting (PGP) Features

IAA production. IAA quantification was determined according to the modified method described by Patten & Glick (2002). Briefly: 150 mL of H_2SO_4 were mixed with 250 mL of distilled water and then 7.5 mL of 0.5 M

$FeCl_3$ were added. Four volumes of reagent were added to one volume of cell-free culture supernatant, mixed gently and, after 20 min in the dark, the absorbance was measured at 535 nm. The concentration of each sample was calculated from a standard curve ranging from 0.01 to 0.4 μg mL^{-1} pure IAA (Sigma).

Phosphorus solubilisation. The ability of the bacterial strains to solubilise tricalcium orthophosphate was tested on PVK medium (Pikovskaya, 1948). 10 μL of bacterial suspensions (about 10^9 CFU mL^{-1}) was dropped onto PVK agar plates. After incubation at 28 °C for 3 days the clearing of the initially turbid medium indicated the presence of phosphate-solubilising strains.

Siderophores production. Semi-quantitative analysis of siderophores production were performed following the chrome azurol S (CAS) method by Alexander & Zuberer (1991). Ten microliters of bacterial inoculum were spotted onto CAS agar plates and incubated at 28 °C for three days; the medium discoloration (from blue to yellow-orange) indicated siderophores production by strains.

Determination of ACC-deaminase activity. ACC deaminase activity was determined according to Penrose & Glick (2003). Bacterial cultures grown overnight (28 °C, 180 rpm) in tryptic soybean broth (TSB) were centrifuged and washed twice with Dworkin and Foster (DF) salts minimal medium without nitrogen source. Then, they were resuspended in DF salts minimal medium with 3.0 mM ACC, which induces ACC deaminase activity, and incubated overnight under the same conditions. After that, cultures were washed twice in 0.1 M Tris-HCl, pH 7.6 and suspended in 0.1 M Tris-HCl, pH 8.5. 30 mL of toluene was added to each cellular suspension to lyse the cell membranes, and the mixture was homogenised using a vortex. Using two tubes for each sample, 20 mL of 0.5 M ACC was added in one of the tubes. Both tubes were incubated at 30 °C for 15 min. Following to the addition of 1 mL of 0.56 M HCl, the mixture was homogenised and centrifuged. One mL of supernatant was mixed with 800 mL of 0.56 mM HCl, and then 300 mL of 0.2 % 2.4-dinitrophenyl-hydrazine in 2 M HCl was added. The mixture was homogenised and incubated at 30 °C for 30 min. 2 mL of 2 M NaOH was added for colour development.

The absorbance was measured at 540 nm. In order to calculate the amount of α-ketobutyrate generated, the absorbance of the tube non-containing ACC was subtracted to the absorbance of the tube containing ACC and then compared with an α-ketobutyrate standard curve.

2.3. Strain Identification

Chromosomal DNA of strains CT42, CT276, CT363 and M16 was isolated using Wizard Genomic DNA Purification Kit (Promega). The 16S rRNA coding gene was amplified by PCR using universal primers 27f and 1492r (Lane, 1991) at an annealing temperature of 55.5 °C. Amplified DNA fragments were cloned using the TOPO TA cloning kit 2.1 (Invitrogen). The sequencing of purified plasmid DNA was performed using BigDye Terminator v.3.1 cycle sequencing kit (Applied Biosystems) as directed by the manufacturer´s protocol and an ABI PRISM 3730 DNA analyser (Applied Biosystems). Sequence data were edited and assembled by BIOEDIT (Sequence Alignment Editor) v.7.0.9 (Hall, 1999), carrying out an alignment with those phylogenetically closely related species (as determined by EzTaxon server) (Chun et al., 2007). The 16S rRNA gene sequences were deposited in DDBJ/EMBL/GenBank database.

2.4. Gnobiotic Assays

Canola root elongation assay. The effect of bacterial strains on root elongation of canola (*Brassica napus*) seeds was assessed according to Patten & Glick (2002). Canola seeds (Thunder) provided by Department of Biology of the University of Waterloo (Ontario, Canada) were disinfected with 70 % ethanol for 1 min and 1 % sodium hypochlorite for 10 min followed of six washes with sterile distilled water. Seeds were submerged in 4 mL of bacterial suspension (final absorbance equals to 0.15 at 600 nm) prepared in sterile 0.03 M $MgSO_4$ or in just sterile 0.03 M $MgSO_4$ (negative control) for one hour. Then, seeds were placed into sterile growth pouches (Mega International, Minneapolis, MN, USA) filled with 12 mL of distilled water and incubated in a growth chamber (Sanyo Electric Co., Japan) at 20-25 °C, 80 % RH and photoperiod of 12 h (4 LS). Each bacterial assay (nine growth pouches with eight treated seeds each) and its accordingly negative control were performed independently. In order to avoid the desiccation, two empty wet growth pouches were placed at the ends of each rack. After 3 days, non-germinated seeds were discarded. Primary root lengths were measured after 7 days.

Rooting assays. Effects of bacterial strains on formation and development of adventitious roots were assessed using cuttings of mung beans *Vigna radiata* L. (Wilczek) provided by Penelas Legumes, S.A. (León, Spain) according to Montero-Calasanz et al. (2013a). To determine effects of different microbial IAA concentrations on plants, additional experiments using *Vigna unguiculata* L. walp. cv. Red Caloona cuttings were carried out according to Mayak et al. (1999) and Tsavkelova et al. (2007) with minor modifications. Mung bean seeds were disinfected with 96 % ethanol for 30 s and 6 % sodium hypochlorite for 6 min followed by six washes with sterile distilled water. Then seeds were placed on water agar plates (1 % agar) at 28°C for 24-48 h in darkness to stimulate seed germination. Seedlings were sowed individually in glass tubes (3 × 25 cm) closed with cotton plugs containing quartz sand and nutrient solution (Rigaud and Puppo, 1975) under sterile conditions. Tubes were maintained in a growth chamber (Sanyo Electric Co., Japan) at 20-25 °C, 80 % RH and photoperiod of 12 h (4 LS). After 1 week, stems were cut above the neck. Cuttings (8-10 cm in length) were managed in four different ways. Negative control and bacterial treatments were placed in glass tubes (2.5 × 15 cm) closed with cotton plugs containing 6 mL sterile distilled water and 6 mL bacterial suspensions (bacterial cultures centrifuged, washed twice with sterile distilled water and resuspended in the same volume), respectively. Positive control and supernatant treatments were nevertheless introduced in 50 ppm IAA solution for 1 h and in the first supernatant got from bacterial culture, respectively, placing them after that in glass tubes closed with cotton plugs containing 6 ml sterile distilled water. Ten cuttings per treatment were incubated during 10 days in a growth chamber under the environmental conditions previously described. After this period, the number and length of adventitious roots were measured.

2.5. Olive Nursery Assays

Rooting assays. Olive semi-hardwood cuttings (12-15 cm long) were cut from vigorous one-year-old shoots of pathogen-free trees (10-15 years old) in Burguillos (Seville, Spain) to carry out four olive rooting assays. Assay 1 was managed in summer using cultivars arbequina, hojiblanca and picual. Assay 2 was performed in spring using ‘Arbequina’. Assays 3 and 4 tested `Arbequina´ and `Koroneiki´ in both mentioned seasons. Assays were carried out in a commercial nursery to take advantage of its propagation scheme. Cuttings were inoculated by submerging the basal end of cuttings (1 cm) for 24 h in a bacterial culture (Assay 1) or for 2 h (Assay 2, 3 and 4). Negative and positive controls were immersed in nursery water for 24 h or 2 h and in an IBA solution (3000 mg l^{-1}) for 5 s, respectively. Cuttings were planted at 2 cm depth in alveolar trays of 130 alveolus (2 cm long × 2 cm wide × 4 cm depth) containing mixed damp coconut fibre and perlite (3:1). After planting, cuttings were placed in a propagation tunnel, built inside of a climatized greenhouse, provided of an air (20 ºC) and bed (22 ºC) electronically controlled temperatures. Intermittent mist (0.3 mm nozzle) was controlled by an electronic device. Three replicates of 78 and 52 cuttings each randomly distributed in the propagation tunnel were tested in assays 1 and 2 and 3 and 4, respectively. After 60 days ('Arbequina', 'Picual' and `Koroneiki´) or 74 days ('Hojiblanca'), rooted cuttings percentages were calculated.

Long-term effect assay. An *assay* to evaluate rooted olive cuttings by PGPB in *long-term* was performed. Twenty-five rooted olive cuttings per bacterial treatment together with positive and negative controls were transplanted to pots (6.5 x 7 cm), containing commercial peat moss, and placed under shadow net. After one year, plants were carefully extracted and washed thoroughly in running tap water until to remove the adhered substrate. Then aerial and root lengths, neck diameter and root and aerial dried weight were determined.

2.6. Statistical Analyses

Results were statistically analysed with SPSS v.15.0 (Lead Technologies, Inc.). Analysis of variance (ANOVA) was performed for gnobiotic mung bean assays and olive nursery assays data, carrying out multiple comparisons between means according to established experimental design. In *Vigna unguiculata* experiments, a factorial ANOVA was additionally applied for an appropriate interpretation of the interaction effects and its profile plots were represented. A Student's t-test was used to assess differences in root length of canola between inoculated and non-inoculated growth pouches.

3. Results

3.1. Strain Characterization

Four new bacterial isolates were characterised for a range of plant growth promoting traits including IAA and siderophores production, phosphorus solubilisation, and ACC deaminase activity (Table 1).

Table 1. Plant growth promoting features and taxonomic affiliation of bacterial strains

Bacterial strain[1]	IAA production[2] (mg/l)	Phosphorus solubilisation[3]	Siderophore production[3]	Acc deaminase activity[4] (nmol α-ketobutirate/mg of protein/h)	Taxonomic affiliation[5]
CT42 (GU124698)	55.1 ± 10.4	+	+	0.0	*Pseudomonas* reinekei Mt-1^T(99.2%)
CT276 (HQ455829)	58.1 ± 8.3	-	-	0.0	*Arthrobacter* oryzae KV-651^T (99.7%)
CT363 (GU124699)	12.1 ± 3.4	++	+++	0.0	*Flavimonas* oryzihabitans IAM 1568^T (99.2%)
M16 (HQ455820)	75.2 ± 14.5	+	+	1270.1 ± 141	*Enterobacter asburidae* JCM 6051^T (99.4%)

[1] DDBJ/EMBL/GenBank accession number in brackets. [2]N=6. Mean ± SD. [3]Based on the diameter of hydrolysis halo formed on agar plates: -, negative; +, 1-2.99 cm; ++, 3-5.99 cm; +++, > 6 cm). [4]ACC: 1-aminocyclopropane 1-carboxylate. [5] 16S rRNA gene sequence similarity in brackets.

All the strains tested were IAA and siderophores producers and phosphorus solubilisers, except the rhizospheric strain CT276 that only synthesised IAA. ACC deaminase activity was only exhibited by strain M16.

The taxonomic affiliation based on 16S rDNA gene sequencing placed strains CT42, CT363 and M16 within the class *Gammaproteobacteria,* specifically in the well-known plant associated bacterial families *Pseudomonadaceae* and *Enterobacteriaceae.* Strain CT276 was identified as a member of *Micrococcaceae,* another known PGPB family and within the genus *Arthrobacter* (Table 1).

3.2. Gnobiotic Assays

Root elongation assay. All tested strains significantly stimulated the growth of canola seedling primary roots after seven days of treatment (Table 2), with increments from 6-13% with respect to negative controls. Plants treated with strain *Enterobacter asburidae* M16 displayed the longest root lengths.

Table 2. Effects of bacterial strains on root length of canola under gnotobiotic conditions

Bacterial treatment	Root length (mm)
CT42	63.7 a
Negative control	59.9 b
CT276	57.4 a
Negative control	53.0 b
CT363	65.4 a
Negative control	60.2 b
M16	67.8 a
Negative control	59.9 b

Values are means of 72 replicates from three independent experiments. Plants were collected at 7 days after inoculation. Each bacterial treatment was compared with its own control. Means followed by different letter are significantly different at $p < 0.05$, based on Student's t-test.

Rooting assays. In *Vigna radiata* experiments, all bacterial treatments increased both number of adventitious roots and total root length from three to seven fold when compared with negative control, showing strain M16 the highest values in number of roots and total root length per cutting (Table 3).

Table 3. Effects of bacterial suspensions on the number and total length of adventitious roots on *Vigna radiata* cuttings under gnotobiotic conditions

Bacterial treatment	No. of roots/cutting	Total root length/ cutting (mm)	Rate of average root length
Control	6.6 f	32.6 d	4.9 bc
CT42	22.4 de	174.0 b	7.8 a
CT276	26.8 cd	129.2 c	4.8 bc
CT363	34.9 ab	189.4 b	5.4 bc
M16	39.6 a	246.0 a	6.2 abc

Values are means of 10 replicates from three independent experiments. Plants were collected at 10 days after inoculation. Means followed by the same letter are not significantly different at $p < 0.05$ by Duncan´s Multiple Range Test. Rate of average root length = average total root length/ average roots number.

It was nevertheless *Pseudomonas reinekei* CT42 suspension, which developed the best rate of average root length (7.8 mm on average).

In *Vigna unguiculata* experiments, most bacterial treatments significantly differed from the negative control in both parameters, highlighting effects induced by strains *Arthrobacter oryzae* CT276 and *Enterobacter asburidae* M16 (Table 4).

Considering the rate of average root length, the best values were nevertheless displayed by *Pseudomonas koreensis* CT364 and the enterobacterial strain. The application of supernatants showed significant differences between most treatments and the negative control, observing that the supernatant produced by CT276 induced higher results than the positive control for the two studied parameters (Table 4).

Taking into account the rate of average root length, the most efficient supernatant was the produced one by strain CT364. Factorial ANOVA analysis disclosed the use of supernatants exceeds, in almost all treatments, to the bacterial suspensions ones regarding number of emerged roots per cutting, mainly supernatants produced by strains CT42, CT276, CT348, CT364 and M16 (Table 4 and Supplementary Figure S1).

Contrarily, it was observed that the application of bacterial suspensions induced the longest roots (Supplementary Figure S2). This trend became significantly different for treatments inoculated by *P. reinekei* CT42, *Flavimonas oryzihabitans* CT363 and *E. asburidae* M16. Treatments inoculated by CT276 however showed results significantly opposed to the others. The study of the rate of average root length confirmed the above results (Supplementary Figure S3).

Table 4. Effects of supernatant and bacterial suspension on the number and total length of adventitious roots on *Vigna unguiculata* cuttings under gnotobiotic conditions

Treatments	Applied method					
	Supernatant			Bacterial suspension		
	No. of roots/cutting	Total Root length/ cutting (mm)	Rate of average root length	No. of roots/cutting	Total Root length/ cutting (mm)	Rate of average root length
Negative	7.6 e A	78.6 f A	10.3 bc A	7.6 d A	78.6 f A	10.3 e A
Positive (IBA)	11.4 cd	127.4 cd	11.1 ab			
AG9	9.9 de A	103.7 def A	10.4 bc B	9.8 bc A	127.0 de A	12.9 de A
AG13	8.2 e A	91.6 ef A	11.2 ab A	10.4 bc A	123.0 e A	11.7 e A
CT42	13.8 bc A	101.2 def B	7.3 bc B	9.1 cd B	160.3 cd A	17.5 bcd A
CT276	53.9 a A	298.7 a A	5.5 c A	29.3 a B	217.4 ab B	7.4 e A
CT348	13.5 bc A	142.1 c A	10.5 b B	10.5 bc B	158.5 cd A	15.0 cd A
CT363	13.0 bcd A	123.8 cde B	9.5 bc B	10.9 bc A	190.2 bc A	17.4 abc A
CT364	12.7 bcd A	205.5 b A	16.2 a B	9.7 bcd B	211.8 ab A	21.9 a A
M16	15.6 b A	127.7 cd B	8.2 bc B	11.6 b B	247.7 a A	21.4 ab A

Values are means of 10 replicates from the three separate experiments. Means followed by the same lower case letter in the same column are not significantly different at $P < 0.05$, based on Duncan´s Multiple Range test. Means followed by the same capital letter in the same line comparing with the same parameter are not significantly different at $P < 0.05$, using a factorial analysis of variance based on Bonferroni´s multiple comparisons method at $P < 0.05$. Rate of average root length = average total root length / average no. roots.

3.3. Olive Nursery Assays

Rooting assay 1. Treatments inoculated with *Pantoea* sp. AG9 and the new isolate *F. oryzihabitans* CT363 were the most effective and versatile, providing similar or even higher percentage of rooted cuttings than the IBA control in the three tested cultivars (Table 5). The similar rooting induction to the IBA control produced by strains M16 and CT364 in `Arbequina´, strains CT42 and CT348$^{T}$ in `Picual´ and strains *Azospirilum brasilense* Cd and M16 in `Hojiblanca´ was also noteworthy.

Table 5. Effect of bacterial treatments on the percentage of rooted cuttings in 'Arbequina', 'Picual' and 'Hojiblanca' (Assay 1)

Treatment	Cultivar		
	Arbequina[1]	Picual[1]	Hojiblanca[1]
Cd	16.7 de	31.3 de	40.0 bcd
AG9	60.5 a	61.5 a	48.5 ab
CT42	22.3 cde	51.8 abc	16.0 ef
CT276	26.5 cd	22.0 e	23.0 def
CT348^{T}	19.5 de	48.3 abc	16.5 def
CT363	46.5 ab	54.5 ab	70.5 a
CT364	38.0 bc	39.0 cd	19.5 def
M16	51.7 ab	40.0 bcd	40.5 bcd
Nursery water	6.7 e	7.7 f	10.0 f
IBA	35.7 bc	48.0 abc	42.7 bc

Values are means of three replicates, 78 cuttings each. Cuttings were inoculated by submerging the basal end of cuttings for 24 h in a bacterial culture. Cuttings were collected at 60 days after inoculation, except cv. hojiblanca that was harvested after 74 days. Means followed by the same letter are not significantly different at $p < 0.05$ by Duncan´s Multiple Range Test.

Rooting assay 2. The reduction of the inoculation time from 24 h to 2 h did not influence on the bacterial rooting induction in cultivar arbequina (Table 6). All of them, except strain AG13^{T}, yielded similar rooting percentages than the IBA control in this cultivar, and these results were comparable to those obtained in previous assay using a longer inoculation time.

Rooting assays 3 and 4. A new cultivar, namely koroneiki, was introduced in these assays owing to its importance in Greece, where it is the predominant oil-producing cultivar (COI, 2013). In both assays, the highest results were obtained by *Chryseobacterium oleae* CT348^{T} that induced higher rooting percentages than 70 % and even close to 90 % (Table 7). Furthermore, all treatments in the cultivar arbequina were as efficient as the IBA control, except the treatment inoculated with *Flavimonas oryzihabitans* CT363 in the second assay. Concerning to the cultivar koroneiki, all bacterial treatments, except the inoculated ones by CT276 and CT363 in the first assay, yielded similar rooted cuttings percentage to the positive control.

Table 6. Effect of bacterial treatments on the percentage of rooted cuttings in 'Arbequina' (Assay 2)

Treatment	cv Arbequina	Treatment	cv Arbequina
Cd	34.5 ab	CT276	45.7 a
AG9	48.3 a	CT348^T	46.5 a
AG13^T	27.3 b	CT363	33,5 ab
CT42	43.5 a	CT364	45.5 a
Nursery Water	20.5 b	M16	47.0 a
IBA	46.0 a		

Values are means of three replicates, 78 cuttings each. Cuttings were inoculated by submerging the basal end of cuttings for 2 h in a bacterial culture. Cuttings were collected at 60 days after inoculation. Means followed by the same letter are not significantly different at $p < 0.05$ by Duncan´s Multiple Range Test.

Table 7. Effect of bacterial treatments on the percentage of rooted cuttings in 'Arbequina' and 'Koroneiki' (Assays 3 and 4)

Treatment	Cultivar			
	Arbequina		Koroneiki	
	Assay 3	Assay 4	Assay 3	Assay 4
AG9	36.0 b	55.4 bc	38.0 b	56.7 bc
CT276	45.6 b	63.9 ab	9.0 de	42.4 c
CT348^T	70.7 a	77.7 a	73.5 a	89.1 a
CT363	38.4 b	42.5 d	16.1 d	52.4 bc
CT364	69.0 a	56.3 abc	54.3 ab	70.6 ab
IBA	49.0 ab	72.4 ab	33.0 bc	74.8 ab

Values are means of three replicates, 52 cuttings each. Cuttings were inoculated by submerging the basal end of cuttings for 2 h in a bacterial culture. Cuttings were collected at 60 days after inoculation. Means followed by the same letter are not significantly different at $p < 0.05$ by Duncan´s Multiple Range Test.

Table 8. Long term effects of rooting induction by PGPB on olive tree cuttings in `Arbequina´

Treatment	Biometric parameters				
	NDiameter (cm)	ADW (g)	APLength (cm)	RDW (g)	RLength (cm)
AG9	3.7 a	1.6 ab	17.5 ab	0.7 a	28.3 a
CT42	3.3 bc	1.2 bc	13.8 bc	0.6 ab	24.5 a
CT363	3.6 ab	1.5 ab	17.0 ab	0.7 a	24.3 a
CT364	3.5 abc	1.7 a	20.2 a	0.7 a	24.2 a
M16	3.3 abc	1.4 ab	16.6 ab	0.6 ab	27.3 a
Nursery water	3.2 c	0.9 c	12.4 c	0.5 b	22.7 a
IBA	3.6 ab	1.4 ab	16.6 ab	0.7 a	25.4 a

Values are means of 25 plants after one year under greenhouse conditions. Means followed by the same letter are not significantly different at $p < 0.05$ by Duncan´s Multiple Range Test. NDiameter, neck diameter; ADW, aerial dried weight; APLength, aerial part length; RDW, root dried weight; RLength, root length.

Long-term effect assay. L*ong term effects* of rooting induction by strains AG9, CT42, CT363, CT364 and M16 on olive tree cuttings from the cultivar arbequina are shown in Table 8. After one year, all the tested parameters were similar to the IBA control, showing there is not a detrimental effect of the bacterial treatment on the plant growth.

4. Discussion

The commercial olive tree propagation requires of specific infrastructures, is time-consuming and prevents working under sterile conditions, complicating the screening of potential PGPB to be used as promoters in rooting induction of olive cuttings. In this work, the validity of phenotypic *in vitro* tests and gnotobiotic assays carried out on highly sensitive to hormonal changes models plants as tools to evaluate putative PGPB was confirmed, in accordance with others studies (Barriuso et al., 2005). It was also supported by the taxonomic affiliation of selected microorganisms, that belonged to genera whose members have been widely described as plant growth promoters (Gray and Smith, 2005; Cheng et al., 2007; Gutiérrez Bustos et al., 2009; Shoebitz et al., 2009).

The root response to plant hormone changes such as auxin levels, may vary depending on origin and anatomical and physiological root characteristics (Mayak et al., 1999). Stimulating effects on primary root of canola carried out by suspensions of PGPB strains confirmed the previous observations of Patten and Glick (2002), Ali et al. (2009) and Montero-Calasanz et al. (2013a) relating to low levels of IAA are able to stimulate the root elongation. As well, bacterial suspensions on mung beans cuttings, a model plant widely used in studies on the involvement of phytohormones (Tsavkelova et al., 2007), revealed a significant stimulation on initiation and development of adventitious roots, confirming data of Li et al. (2005) and Montero-Calasanz et al. (2013a) for other PGPB. The highly IAA producer *Enterobacter asburidae* M16 got the best results. This fact let to establish a positive correlation between *in vitro* IAA production and increments in root weight, in agreement with reported by Khalid et al. (2004). Arkhipova et al. (2005) explained it due to increments in the endogenous concentration of IAA in plant caused by bacterial auxin production. Nevertheless, the highest rate of average root length displayed by strains *Pseudomonas reinekei* CT42 and *P.* sp. CT364 suggests that highly IAA producer strains, such as M16 and CT276, might activate the enzyme ACC synthase in plant, raise the ethylene levels and subsequently produce smaller roots (Mayak et al., 1999; Penrose and Glick, 2003).

This hypothesis would be strongly supported by the results obtained in the supernatant treatments of *Vigna unguiculata* assays and especially by *Arthrobacter oryzae* CT276, which could have overexposed the plants to IAA, taking place a high roots number per cutting but also turning on the elongation inhibition. However, *a priori* the presence of ACC deaminase activity by strain M16 would be buffering the increasing ethylene levels, at least partially, not blocking the root elongation and therefore, yielding a large number of moderate adventitious roots. This fact, in a prospective way, would improve plant health and survival rates of seedlings in commercial nurseries in front of others.

In recent years the popularity and use of liquid formulations has enhanced based on easy application either on the seed or *in situ* delivery (Xavier et al., 2004), equivalent results to peat-based inoculants (Kellman, 2008) and lacking of contaminants (Benintende, 2010), as well as by solving the problems associated with processing solid carriers (Atieno et al., 2012). Attending to these facts, here we optimised our already described method of propagation of olive tree cuttings, lowering the dipping time in the bacterial culture. The diminution of the inoculation time from 24 h to 2 h did not influenced on the bacterial adherence (data not shown) and therefore, on the rooting induction in the cultivar arbequina, providing a high effectiveness in nursery handling without loss in production.

The bacterial inoculation conducted to similar or significantly higher rooting percentages than positive control in the four assayed olive tree cultivars. It was mainly highlighted the rooting potential and adaptability of strain AG9 in all the tested cultivars, confirming the previous observations of Montero-Calasanz et al. (2013a). In addition, it was worthy the rooting close to 90 % carried out by *Chryseobacterium oleae* CT348^{T} in `Koroneiki´, a cultivar characterised by a low-medium rooting ability (COI, 2013).

Given that the strain CT348^{T} is not the best IAA producer and lacks of enzyme ACC deaminase, its effectiveness could be explained by other mechanisms not taking into account in the *in vitro* tests, together with the specificity to root exudates from this cultivar. Likewise, it was observed that some effective strains with one cultivar produced a low rooting percentage with others. This different behaviour, also observed between the two *Vigna* spp., confirmed the complex interaction between both partners (plant and microorganism) described previously (Montero-Calasanz et al., 2013a).

Adventitious roots induction by high levels of IAA depends on the vegetal species (Patten and Glick, 2002; Ali et al., 2009) and is the base of the nursery industry. The application in the base of olive cuttings of a longer IBA dipping time, a frequent problem, since it is the nurseryman´s experience who determines it, causes important losses in the crop production. The use of PGPB avoids the overexposure, since cuttings are subjected to comparatively low auxin concentrations (10-100 ppm) during all the mist period (data not shown), therefore losses are nonexistent, could also reach a highlighted importance in the conventional nursery production.

The olive tree propagation method by PGPB under real nursery conditions confirmed here, is effective, efficient, ecological and plant harmless in a long-term in contrast with other propagation methods and could spread out into cultivars showing low rooting ability and compete directly with the conventional propagation method. Nevertheless, this technology still needs additional studies about bacterial survival and colonization in the system in order to optimise the development of commercial inoculants that ensure successful rooting results.

ACKNOWLEDGMENT

The authors thank University of S. Pablo CEU (Madrid) for providing strain M16. This research was supported by IFAPA-DGAE Project Exp. DAP 92162 nº 65. M.C. Montero-Calasanz was the recipient of a grant from INIA.

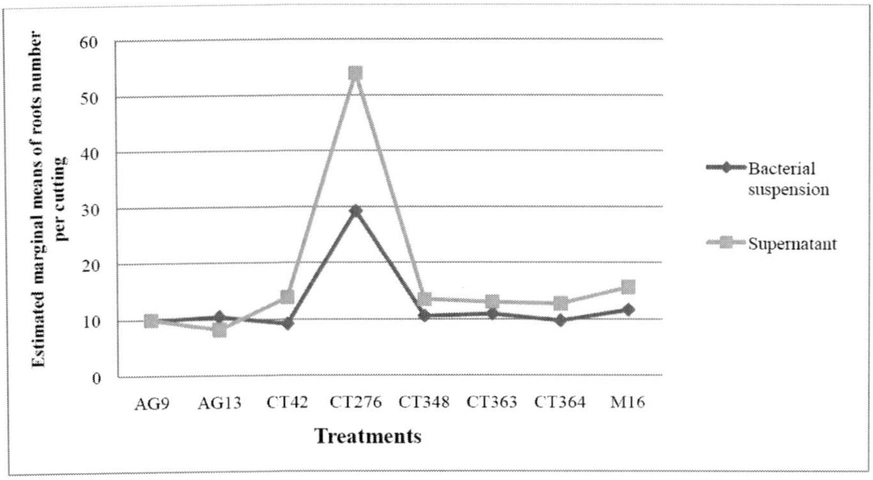

Suplementary Figure S1. Estimated marginal means profile plots of root number per *Vigna unguiculata* cutting for the two applied methods.

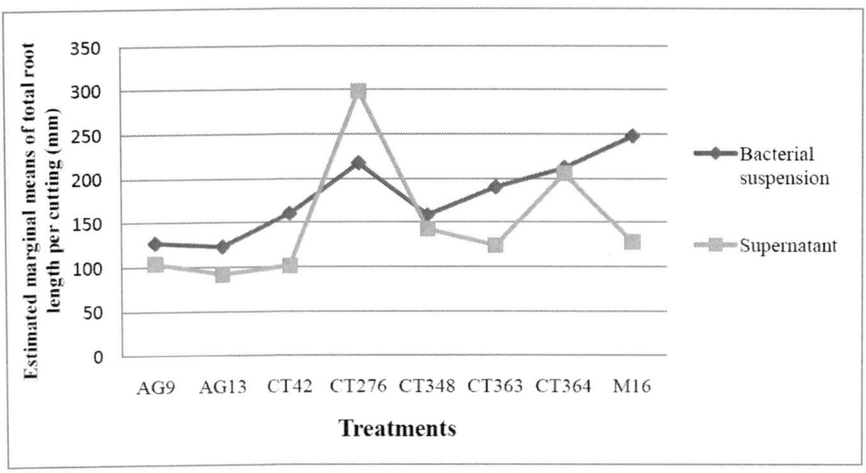

Suplementary Figure S2. Estimated marginal means profile plots of total root length per *Vigna unguiculata* cutting for the two applied methods.

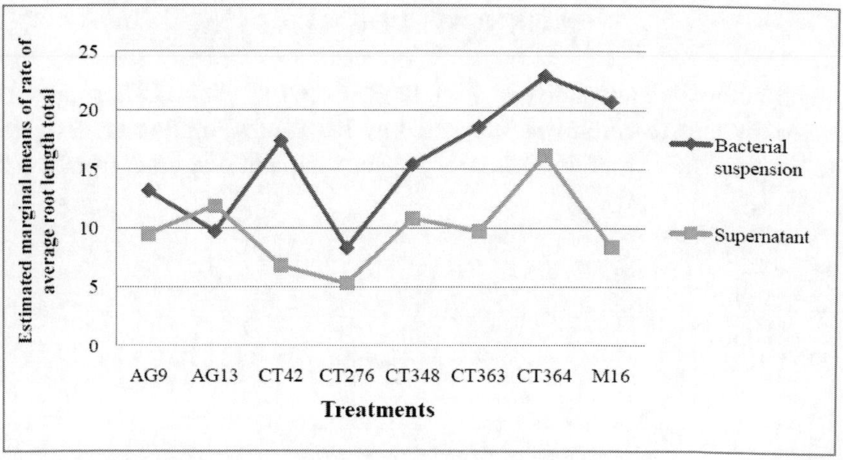

Suplementary Figure S3. Estimated marginal means profile plots of average total root length of *Vigna unguiculata* cuttings for the two applied methods.

REFERENCES

Abbas-Zadeh P, Saleh-Rastin N, Asadi-Rahmani H, Khavazi K, Soltani A, Shoary-Nejati AR, Miransari M. Plant growth-promoting activities of fluorescent pseudomonads isolated from the Iraian soils. Acta *Physiol Plant* 2010; 32: 328-288.

Ali B, Sabri AN, Ljung K, Hasnain S. Quantification of indole-3-acetic acid from plant associated Bacillus spp. and their phytostimulatory effect on *Vigna radiata* (L.). *World J Microbiol Biot* 2009; 25: 519-526.

Arkhipova TN, Veselov SU, Melentiev AI, Martynenko EV, Kudoyarova GR. Ability of bacterium *Bacillus subtilis* to produce cytokinins and to influence the growth and endogenous hormone content of lettuce plants. *Plant Soil,* 2005; 272: 201-209.

Atieno M, Herrman L, Okalebo R, Lesueur D. Efficieny of different formulations of *Bradyrhizobium japonicum* and effect of co-inoculation of *Bacillus subtilis* with two different strains of *Bradyrhizobium japonicum. World J Biotechnol* 2012; 28:2541-2550.

Barriuso J, Pereyra MT, García JAL, Megías M, Mañero FJG, Ramos-Solano B. Screening for putative PGPR to improve establishment of the symbiosis *Lactarius deliciosus-Pinus* sp. *Microb Ecol* 2005; 50: 82-89.

Benintende S. Quality of commercial inoculants for soybean crop in Argentina: concentration of viable rhizobia and presence of contaminants. *Rev Argent Microbiol* 2010; 42(2): 129-132.

Cheng Z, Park E, Glick BR. 1-Aminocyclopropane-1-carboxylate deaminase from *Pseudomonas putida* UW4 facilitates the growth of canola in the presence of salt. *Can J Microbiol* 2007; 53: 912-918.

Chun J, Lee J-H, Jung Y, Kim M, Kim S, Kim B K, Lim Y-W. EzTaxon: a web-based tool for the identification of prokaryotes based on 16S ribosomal RNA gene sequences. *Int J Syst Evol Microbiol* 2007; 57: 2259–2261.

Council Regulation (EC) 834/2007 on "Organic Production and labelling of organics products" and repealing Regulation (EEC) 2092/91.

Commission Regulation (EC) 889/2008 on "detailed rules for the implementation of Council Regulation (EC) No 834/2007 on organic production and labelling of organic products with regard to organic production, labelling and control".

Döbereiner J. Forage grasses and grain crops. In: Bergensen FJ, editor. *Methods for Evaluating Biological Nitrogen*. New York: John Wiley and Sons; 1980. p. 535-555.

Glick BR. Using soil bacteria to facilitate phytoremediation. *Biotechnol Adv* 2010; 28: 367-374.

Glick BR, Penrose DM, Li J. A model for the lowering of plant ethylene concentrations by plant-growth-promoting bacteria. *J Theor Biol* 1998; 190: 63-68.

Gray RJ, Smith DL. Intracellular and extracellular PGPR: commonalities and distinctions in the plant-bacterium signaling processes. *Soil Biol Biochem* 2005; 37: 395-412.

Gutiérrez Bustos DC, Hernández Marín AM, Corrales Ramírez JC. *Pseudomonas Oryzihabitans*: Un microorganismo de creciente interés científico. Nova. Publicación Científica en Ciencias Biomédicas 2009; 7 (11): 103-112.

Hall TA. BioEdit: a user-friendly biological sequence alignment editor and analysis program for Windows 95/98/NT. *Nucl Acids Sym Ser* 1999; 41:95-98.

International Olive Oil Council (COI). 2012; (http://www.international oliveoil.org/).

Kellman AW. Rhizobium inoculation, cultivar and management effects on the growth, development and yield of common bean (*Phaseolus vulgaris* L.).

PhD Thesis. New Zealand: Lincoln University Canterbury; 2008. p. 189-190.

Khalid A, Arshad M, Zahir ZA. Screening plant growth-promoting rhizobacteria for improving growth and yield of wheat. *J Appl Microbiol* 2004; 96: 473-480.

Lane DJ. 16S/23S rRNA sequencing. In: Stackebrandt E, Goodfellow M, editors. *Nucleic acid techniques in bacterial systematics*. New York: John Wiley and Sons; 1991. p. 115-175.

Li Q, Saleh-Lakha S, Glick BR. The effect of native and ACC deaminase-containing *Azospirillum brasilense* Cd1843 on the rooting of carnation cuttings. *Can J Microbiol* 2005; 51: 511-514.

Mayak S, Tirosh T, Glick BR. Effect of wild-type and mutant plant growth-promoting rhizobacteria on the rooting of mung bean cuttings. *J Plant Growth Regul* 1999; 18: 49-53.

Montero-Calasanz MC, Göker M, Rohde M, Spröer C, Schumann P, Busse H-J, Schmid M, Tindall BJ, Klenk H-P, Camacho M. *Chryseobacterium oleae* sp. nov., an efficient plant growth promoting bacterium in the rooting induction of olive tree (*Olea europaea* L.) cuttings and emended descriptions of the genus *Chryseobacterium, C. daecheongense, C. gambrini, C. gleum, C. joostei, C. jejuense, C. luteum, C. shigense, C. taiwanense, C. ureilyticum* and *C. vrystaatense*. *Syst Appl Microbiol.* 2014. doi: 10.1016/j.syapm.2014.04.004.

Montero-Calasanz MC, Göker M, Rohde M, Spröer C, Schumann P, Busse H-J, Schmid M, Tindall BJ, Klenk H-P, Camacho M. *Chryseobacterium hispalense* sp. nov., a plant growth-promoting bacterium isolated from a rainwater pond in an olive plant nursery and emendation of the species *Chryseobacterium defluvii, Chryseobacterium indologenes, Chryseobacterium wanjuense* and *Chryseobacterium gregarium*. *Int J Syst Evol Microbiol* 2013b; 63:4386-4395.

Montero-Calasanz MC, Santamaría C, Albareda M, Daza A, Duan J, Glick BR, Camacho M. Alternative rooting induction of semi-hardwood olive cuttings by several auxin-producing bacteria for organic agriculture systems. *Span J Agric Res* 2013a; 11: 146-154.

Parra López C, Calatrava Requena J. Análisis de factores relacionados con la adopción de la forma de producción ecológica en el olivar del sur de España. Prod. I IFOAM Worldwide Conference on Organic Olive Farming, Sierra de Génave (Jaén), Spain, May 22-24, 2002.

Patten CL, Glick BR. Role of *Pseudomonas putida* indolacetic acid in development of the host plant root system. *Appl Environ Microbiol* 2002; 68: 3795-3801.

Penrose DM, Moffat BA, Glick BR. Determination of 1-aminocycopropane-1-carboxylic acid (ACC) to assess the effects of ACC deaminase-containing bacteria on roots of canola seedlings. *Can J Microbiol* 2001; 47: 77-80.

Penrose DM, Glick BR. Methods for isolating and characterizing ACC deaminase-containing plant growth-promoting rhizobacteria. *Physiol Plant* 2003; 118: 10-15.

Ramos-Solano B, Lucas JA, García-Villaraco EA, García-Cristobal J, Gutiérrez-Mañero FJ. Siderophore and chitinase producing isolates from the rhizosphere of *Nicotiana glauca* Graham enhance growth and induce systemic resistance in *Solanum lycopersicum* L. *Plant Soil* 2010; 334: 189-197.

Rigaud J, Puppo A. Indole-3-acetic-acid catalism by soybean bacteroids. J Gen Microbiol 1975; 88: 223-228.

Shoebitz M, Ribaudo CM, Pardo MA, Cantore ML, Ciampi L, Curá JA. Plant growth promoting properties of a strain of *Enterobacter ludwigii* isolated from *Lolium perenne* rhizosphere. *Soil Biol Biochem* 2009; 41 (9): 1768-1774.

Spaepen S, Vanderleyden J, Remans R. Indole-3-acetic acid in microbial and microorganism-plant signalling. *FEMS Microbiol Rev* 2007; 31: 425-448.

Toklikishvili N, Dandurishvili N, Vainstein A, Tediashvili M, Giorgobiani N, Lurie S, Szegedi E, Glick BR, Chernin L. Inhibitory effect of ACC deaminasa-producing bacteria on crown gall formation in tomato plants infected by *Agrobacterium tumefaciens* or A. vitis. *Plant Pathol* 2010; 59:1023-1030.

Tsavkelova EA, Cherdyntseva TA, Botina SG, Netrusov AI. Bacteria associated with orchid roots and microbial production of auxin. *Microbiol Res* 2007; 162: 69-76.

U.S.D.A- United States Department of Agriculture, Agricultural Research Service. 2013; http://www.usda.gov/wps/portal/usda/usdahome.

Xavier IJ, Holloway G, Leggett M. Development of rhizobial inoculant formulations. Online. *Crop Manag*. 2004. doi:10.1094/CM-2004-0301-06-RV.

In: Auxin
Editor: Rosario D. Pacovsky

ISBN: 978-1-61122-060-5

Chapter 2

AUXIN – ROLE IN PLANT DEFENSE

Magda Aparecida Beneventi[1], Maria Eugênia Lisei de Sá[1,2] and Maria Fatima Grossi–de–Sa[1,3]

[1]Embrapa Genetic Resources and Biotechnology, Brasilia, Brazil
[2]Agricultural Research Company of Minas Gerais State, Uberaba, Brazil
[3]Catholic University of Brasilia, Brasilia, DF, Brazil

ABSTRACT

Auxin is a well-known phytohormone that plays important roles in several physiological processes in plants. Currently, many studies have indicated that multiple hormonal signals, including auxins, are implicated in plant defense signaling pathways, but the roles of these signals in plant defense remain unclear. These phytohormones regulate the gene expression of plant defense and eventually trigger the production of defense molecules, such as phenylpropanoids, phytoalexins and pathogenicity-like proteins (PR). Studies of defense signaling against root-knot nematode invasion in resistant soybean genotype suggest that auxin and gibberellin interactions modulate ROS levels and include the participation of DELLA-like proteins, thereby possibly controlling plant immune and stress responses. Auxins promote the degradation of a family of transcriptional repressors known as Auxin/Indol-3-acetic acid (Auxin/IAA). These repressors bind to auxin repressor factors (ARFs) and inhibit the transcriptional-specific auxin response genes. Most of the total auxin in plants is found in the conjugated form and the formation of auxin conjugates is an important regulatory mechanism in the activation and inactivation of IAA. To regulate plant growth and development, auxin induces the expression of three groups of genes: Aux/IAA, *GH3* and small auxin-up RNA (SAUR) families. Auxin responsive *GH3* genes have been shown to play important roles in plant defense response in *Arabidopsis thaliana* and rice. Recently, it was shown that GH3.5 acts as a bifunctional modulator in both SA (salicylic acid) and auxin signaling during pathogen infection. *A. thaliana* auxin mutants evaluated against *Pseudomonas syringae* had compromised systemic immunity, demonstrating that indole-derived compounds, but not auxin itself, are involved in the maintenance of the systemic immunity. Furthermore, it is suggested that the repression of auxin signaling through either mutations in the auxin pathway or pharmacological interference with the auxin response impairs resistance to the necrotrophic fungi *Plectosphaerella cucumerina* and *Botrytis cinerea*. This chapter describes the influence of auxin on plant responses to stress and defense reactions, including the establishment of effective immunity in the plant-pathogen interaction and the interplay of auxin and other plant hormones.

INTRODUCTION

Auxins are well known phytohormones that affect many plant processes involved in growth, development, and adaptation to environmental conditions. These processes include apical dominance, root gravitropism, and the development of root hairs, lateral roots, leaves, flowers, and plant vasculature (Bari, 2009; Spaepen and Vanderleyden, 2011; Kazan, 2013). Auxins become biologically active in plants through three general biosynthetic routes: indole-3-acetic acid (IAA) biosynthesis and metabolism, cell-to-cell transport of IAA, and SCF^{TIR1}-mediated signaling events that elicit an auxin response (Quint et al., 2009).

IAA is the most studied naturally occurring active auxin and it can be synthesized *de novo* using tryptophan (Trp) as a precursor or via a Trp-independent pathway (Zhao, 2001). Several Trp-dependent auxin biosynthesis pathways contribute to IAA levels, including the IAOx (indole-3-acetaldoxime) pathway, the IAM (indoleacetamine) pathway, and the IPyA (indole-3-pyruvic acid) pathway (Korasick et al., 2013). Auxin binding to its receptor F-box protein TIR1 (a subunit of the ubiquitin ligase complex SCF^{TIR1}) stabilizes the interaction with a family of transcription repressors (auxin/indole acetic acid (Aux/IAA)).

Thus Aux/IAA proteins are degraded by SCF^{TIR1} ubiquitin complex and relieves the repression of auxin response factors resulting in auxin-responsive genes transcription (Benjamins and Scheres, 2008; Mockaitis and Estelle, 2008; Santner and Estelle, 2009).

Auxins are involved in a complex network of plant–pathogen interactions (Ghanashyam and Jain, 2009), but the relationship between plant defense and auxin signaling are still poorly understood (Jones and Dangl, 2006). It has been shown that pathogen infection results in auxin imbalance and altered gene expression (Bari, 2009). Some evidence suggests a role of auxin in plant defense responses and crosstalk among the auxin, biotic and abiotic stress signaling pathways (Ding et al., 2008). Auxin regulates these developmental processes by altering the expression of several genes, including the primary auxin-responsive genes of auxin/indole-3-acetic acid (Aux/IAA), the small auxin-upRNA (SAUR), and the GH3 gene families (Ghanashyam and Jain, 2009). Indole-3-acetic acid (IAA) is the most common form of auxin in plants, and it can also be produced by microbes (Fu and Wang, 2011).

GH3 belongs to a multi-gene family that encodes proteins with acyl acid amido synthetase function (Woodward and Bartel, 2005; Westfall et al., 2010) and the transcription of *GH3* genes can be regulated by ARFs (Ulmasov et al.,1999). The functions of the SAUR family are less understood than those of the other two families (Ding et al., 2008). The expression of most SAUR, GH3, and Aux/IAA genes is regulated by auxin response factors (ARFs), which can either activate or repress target gene expression (Woodward and Bartel, 2005).

Many plant pathogenic microorganisms can produce indole-3-acetic acid (IAA), which can be an important component of their pathogenicity (Zhang et al, 2008). Some plant pathogens either produce auxin themselves or increase auxin biosynthesis by the host plant upon infection to manipulate host developmental processes (Manulis et al., 1998; Glickmann et al., 1998; Maor et al., 2004; Vandeputte et al., 2005). Conversely, depending on the type of plant–pathogen association, plants develop either negative or positive regulation of auxin signaling to overcome infection. Thus, this chapter attempts to highlight the strategies used by pathogens to counter host plant immunity and the alternative tactics initiated by the plant during compatible and incompatible interactions that limit pathogen colonization.

The Role of Auxin in Plant-Pathogen Compatible and Incompatible Interactions and Crosstalk with Abiotic Stress

Increasing knowledge of auxin signaling and the metabolic processes involved in plant responses to pathogen invasion has lead to hypotheses of the putative mechanisms of IAA in plant susceptibility and resistance (Table 1). Although most studies focus on its role during pathogenesis, new insights have also been gained into the crosstalk between the biotic and abiotic signaling pathways.

Studies of the complex auxin signal transduction network in plants have been conducted to elucidate the molecular mechanisms underlying the modulation of disease and resistance. Previous studies have reported that auxin is implicated in disease susceptibility and that miRNA-mediated suppression of auxin signaling restricts *P. syringae* growth in *A. thaliana* (Navarro et al., 2006). The Arabidopsis mRNA (miR393a) induced by the flagellin-derived peptide flg22 (TFlg22) targets auxin receptors, thereby contributing to the down-regulation of auxin and leading to the enhancement of *P. syringae* resistance. These findings show that repression of auxin signaling is part of the bacterial-induced plant immune response. Bacterial infection induces the accumulation of indole-3-acetic acid (IAA), which in turn induces the expression of expansins, proteins responsible for cell wall loosening. This mechanism is very important for plant growth but can also make the plant vulnerable to biotic microorganisms (Bari and Jones, 2009). The overexpression of GH3-8, an IAA–amino synthetase that prevents free IAA accumulation, has been shown to activate resistance to the rice pathogen *Xanthomonas oryzae* pv. *oryzae* and inhibit abnormal plant development, likely through the inhibition of expansin expression (Ding et al., 2008). The dual roles of GH3-8 also indicate that it is one of the components of the crosstalk between development and disease resistance pathways, which may partly explain the fitness cost of disease resistance (Figure 1A).

Similarly, it has been demonstrated that in this signaling network, the endogenous auxin content is coordinately regulated through negative feedback by GH3 genes that encode auxin-conjugating enzymes. A mutant auxin-resistant Arabidopsis wes1-D overexpressing a GH3 gene WES1 exhibited typical auxin-deficient traits, such as reduced growth and leaf curling; but it also exhibited resistance to both biotic and abiotic stresses (Park, 2007).

Table 1. Auxin-mediated responses during plant-pathogen interactions

Plant species	Plant/phatogen association	Plant tissue	Pathogen	Time course analysis	Approach	Main genes analysed	Reference
Arabidopsis thaliana	Resistance and susceptibility	Shoot	*Pseudomonas syringae*	30 min with either 10 mM flg22 (+) or 10 mM flg22$^{A.tum}$ (-)	Mutants	F-box auxin receptors: TIR1, AFB2, and AFB3	Navarro et al. (2006)
Arabidopsis thaliana	Resistance and susceptibility	Shoot	*Plectosphaerella cucumerina, Botrytis cinerea, Hyaloperonospora parasitica*	0 to 14 d.a.i	Mutants	AXR1, AXR6, and AXR2, SGT1, RAR1 and RPP4	Llorenrte et al. (2008)
Oryza sativa (Minghui 63)	Resistance	Shoot	*Xanthomonas oryzae pv oryzae (Xoo)*	2 w.a.i	Overexpression	GH3-8, EXPA1, EXPA5 and EXPA10	Ding et al. (2008)
Arabidopsis thaliana	Resistance	Shoot	*Pseudomonas syringae*	2, 4 and 11 d.a.i	Mutants	GH3 (wes1-D)	Park et al. (2007)
Arabidopsis thaliana	Resistance	Shoot	*Psm ES4326 or H. parasitica Noco2*	4 and 8 d.a.i	Mutants	AtMKK7	Zhang et al. (2008)
Oryza sativa (Nipponbare)	Resistance	Shoot	*Magnaporthe grisea*	3 and 4 d.a.i	Microarray	GH3, Aux/IAA, SAUR and ARF	Ghanashyam and Jain (2009)
Oryza sativa (IAC165)	Susceptibility	Root	*Striga hermonthica*	2, 4 and 11 d.a.i	Microarray	GH3, Aux/IAA, SAUR and ARF	Ghanashyam and Jain (2009)
Glycine max (PI 595099)	Resistance	Root	*Meloidogyne javanica*	1, 2, 4, 6 and 8 d.a.i	454 and Sanger sequencing	DELLA-like, auxin and gibberellin-regulated genes	Beneventi et al, (2013); Sá et al. (2012)
Arabidopsis thaliana	Susceptibility	Shoot	*Phytophthora cinnamomi*	3 d.a.i	Mutants	PHR1, f AUX1, AXR1, AXR2 and SGT1B	Eshraghi et al. (2014)

Suppression of auxin signaling by activation of a GH3-type protein enhances bacterial resistance by activating the expression of the PR-1 gene, which is an indicator of systemic acquired resistance in Arabidopsis (Durrant and Dong, 2004).

Antagonistic hormone interaction was demonstrated between MAP kinase kinase 7 (*At*MKK7) gene and auxin signaling during pathogenesis (Zhang et al., 2008). The increased expression of *At*MKK7 in *bud1*-mutants resulted in a significant reduction in free auxin (indole-3-acetic acid) levels in the plants, whereas reducing mRNA levels of *At*MKK7 by antisense RNA expression led to the enhancement of auxin transport, indicating that *At*MKK7 negatively regulates polar auxin transport (PAT) (Dai et al., 2006). The *bud1*-mutant plants accumulated elevated levels of SA (salicylic acid), exhibited constitutive pathogenesis-related (PR) gene expression and enhanced resistance to both the bacterial pathogen *P. syringae* pv. *maculicola* and the oomycete pathogen *Hyaloperonospora parasitica*, suggesting that *At*MKK7 may positively regulate SA signaling and negatively regulate auxin signaling.

The complex crosstalk and hormonal changes that modulate disease and resistance are dependent on pathogen lifestyles and the host genetic constitution (Robert-Seilaniantz et al., 2011). Many microbes involved in symbiotic or pathogenic interactions can synthetize phytohormones that trigger symptoms related to hormonal imbalances (Robert-Seilaniantz et al., 2007; Gimenez-Ibanez and Solano, 2013), but the roles of hormones during plant-pathogen interactions are not fully understood (Bari and Jones, 2009). Similar to plants, some microorganisms produce IAA through multiple metabolic pathways, which routes are classified in terms of their intermediates: the indole-3-acetamide (IAM), the indole-3-pyruvic acid (IPyA), indole-3-acetonitrile, and tryptamine pathways (Berrocal-Lobo and Molina, 2004). The non-gall-forming phytopathogen *Erwinia chrysanthemi* possesses the IAA biosynthesis gene homologues *iaaM* (tryptophan-2-monooxygenase) and *iaaH* (indole-3-acetamide hydrolase) and can produce IAA through the IAM pathway. The null knockout *iaaM* mutant strain showed reduction of IAA production, but its proliferation in *Saintpaulia ionantha* was unaffected. Consequently, the pathogenic effectors, *dspE* (a putative T3SS effector) and *hrpN* (T3SS harpin) genes, belonging to the transcription type III secretion system (T3SS), were impaired in the *iaaM* mutant, suggesting that IAA biosynthesis influences T3SS gene expression (Yang et al., 2007).

In *Gypsophila*, the gall-inducing *Erwinia herbicola* pv. *gypsophilae* harbors both IAM and IPyA pathways for IAA synthesis, as well as cytokinin (CK) biosynthetic genes (Manulis et al., 1998). The inactivation of either *iaaH*, abolishing IAA biosynthesis via the IAM pathway, or *etz*, abolishing cytokinin biosynthesis, produced significantly smaller galls than the wild type (reduction of 37 to 41% in gall size), but still could induce gall formation. In contrast, disruption of the IPyA pathway (*IpdC–* mutant) was found to produce galls of almost equal size as those incited by the parental strain but reduced the pathogen population. Neither the inactivation of both pathways for IAA biosynthesis (*IaaH–, IpdC–*) nor both the IAM pathway and cytokinin biosynthesis (*IaaH–, IaaM–, Etz–*) nor all three pathways (*IaaH–, IaaM–, IpdC–, Etz–)* caused any further significant reduction in gall size greater than that of *IaaH–* or *Etz–* mutants alone. Moreover, a mutant lacking cytokinin production in the absence of both pathways for IAA biosynthesis (*IaaH–, IaaM–, Etz–, IpdC–*) caused slightly less galling than mutants deficient in cytokinin and IAA production via the IAM route alone. These reports suggested that the IAM pathway is involved primarily in gall and further support the view that the IPyA pathway contributes extensively to the epiphytic fitness of this bacterial species.

The involvement of plant hormones in *Plasmodiophora brassicae* development, the causal agent of clubroot disease in brassicas, is supported by the up-regulation of auxin genes related to homeostasis (nitrilases and members of the GH3 family) and down-regulation of cytokinin homeostasis genes (cytokinin synthases and cytokinin oxidases/dehydrogenases) (Siemens et al., 2006). Cytokinin oxidase/dehydrogenase overexpressing lines have been shown to be disease resistant, indicating the importance of cytokinin in clubroot disease development. The *Xanthomonas campestris* pv. *campestris* TTE *AvrBs3* modulates auxin responsive genes, leading to cell hypertrophy (Marois et al., 2002). Sequence analysis revealed homologies to auxin-induced and expansin-like genes, which play a role in cell enlargement. These data suggest that some *AvrBs3*-induced genes may be involved in hypertrophy development and that xanthomonas possess type III effectors in promoting *P. syringae* (biotrophic pathogen) resistance in Arabidopsis (Navarro et al., 2006), as well as another biotrophic parasites (Fu and Wang, 2011). In contrast, there is evidence that SCF-mediated ubiquitination and auxin signaling play a role in promoting resistance to fungal necrotrophs (Llorent et al., 2008). Auxin signaling mutants *axr1*, *axr2*, and *axr6*, which are impaired in auxin-mediated AUX/IAA degradation through SCF^{TIR1}-targeted ubiquitination, exhibited increased susceptibility to the necrotrophic fungi *Plectosphaerella cucumerina* and *Botrytis cinerea*. This suggests that during necrotroph infection of wild-type plants, a moderate suppression of the auxin response is countered by endogenous promotion of the auxin defense that limits pathogen colonization (Figure 1B).

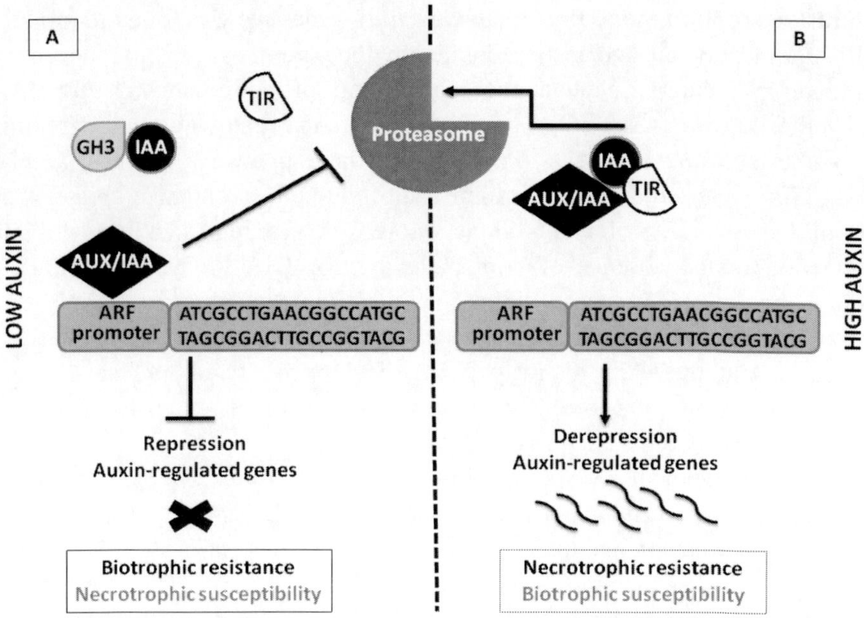

Figure 1. Proposed model of auxin signaling during plant–biotrophic/necrotrophic pathogen interaction. This model explains the integration of the auxin hormonal signal network and plant stress response during pathogen infection. (A) GH3 genes encode IAA-amido synthetases that are involved in the regulation of auxin homeostasis by conjugating excess IAA to amino acids. Auxin promotes the degradation of protein repressors Aux/IAA, which bind to auxin response factors (ARFs) and inhibit the transcription of specific auxin response genes. Thus, low auxin levels result in biotrophic resistance and necrotrophic susceptibility pathogens. (B) The auxin receptor TIR1 (transport inhibitor response 1) bonded to IAA interacts with Aux/IAA proteins. This complex leads to the degradation of Aux/IAA proteins via ubiquitin/26S proteasome pathway and allows the activation of ARFs, resulting in auxin-responsive genes expression. The high IAA level promotes resistance to necrotrophic and disease development for biotrophic pathogens.

Furthermore, *SGT1b* (one of the two Arabidopsis *SGT1* genes encoding HSP90/HSC70 co-chaperones) is known to promote dual functions; e.g., SCF E3-ubiquitin ligase complexes in auxin response, and/or resistance conditioned by particular resistance (R) genes to biotrophic pathogens. It was also demonstrated that *sgt1b* mutants did not increase susceptibility to the necrotrophs *P. cucumerina* due to disruption of auxin pathway (Llorent at al.,

2008). Probably, *sgt1b* defects in SCF E3-ubiquitin ligase functions were also compensated by the presence of SGT1a, which has also *SGT1* activity on ubiquitination process (Azevedo et al., 2006). Conversely, auxin/SCF signaling mutants are uncompromised in their RPP4-triggered resistance to *Hyaloperonospora parasitica* (biotrophic oomycete) and in *sgt1b* mutant plants it was correlated with increased leaf colonization. The predominant action of *SGT1b* in R gene-conditioned resistance to oomycetes appears to occur at a site that does not involve assisting SCF E3-ubiquitin ligases. Therefore, the partial loss of RPP4 resistance observed in *sgt1b* mutants to the obligate biotrophic pathogen *H. parasitica*, combined with no effect of *axr1*, *axr2*, and *axr6* in resistance to this pathogen, and an opposite trend in these mutants in response to *P. cucumerina*, suggest that different mechanisms are engaged in defense responses to biotrophs and necrotrophs. Recent studies have also shown that intact auxin signaling is indispensable for resistance against necrotrophic pathogens (Fu and Wang, 2011).

Crosstalk between plant hormones, primarily gibberellins (GA) and auxins, was recently proposed to explain the positive effect of auxin signaling in the incompatible interaction. This crosstalk may be crucial for modulating the ROS (reactive oxygen species) levels in the resistance reaction to nematode invasion (Sá et al., 2012; Beneventi et al., 2013). Transcriptome analysis of soybean roots infected with *M. javanica* indicated an association with auxin metabolic processes (GO:0009850), as well as the induction of genes encoding proteins that are responsive to the hormone auxin, containing Auxin-inducible, GH3 and AUX/IAA domains. The soybean resistant line PI 595099 appears to encompasses a large diversity of gene families that are important players in parasitic biotrophic-root interaction. These interactions involve multiple hormonal signals, mainly auxins and gibberellins, which may be involved in the coordinated regulation of biological processes that moderate the stress responses and defense reactions against RKNs (root knot nematodes). A plausible model was suggested regarding to the complex mechanisms of crosstalk between gibberellin and other plant hormones, mainly auxins. In brief, upon pathogen infection, jasmonates (JA and MeJA) stimulate auxin biosynthesis while auxin enhances ROS signaled induction. The elevation of ROS levels, in turn, is accompanied by the induction of JAZ/TIFY class of repressors that contribute to prevent jasmonate biosynthesis. Then, auxin elevated levels affect gibberellin biosynthesis leading to induction of gene expression levels of both GA biosynthetic and deactivating genes. Finally, a DELLA-like protein, part of GA-pathway is induced upon pathogen attack. It is proposed to act as a key element to integrate signals from auxin, GA and JA in both growth and defense response processes mainly causing ROS levels to remain low and positively regulating anthocyanin accumulation and PCs (phenolic compounds) as part of the redox system under stress conditions. This model is presented in more detail and illustrated in Beneventi et al. (2013).

Recently, the interplay of auxin between biotic and abiotic stresses was reported, involving a range of hormonal mechanisms that act together in a complex regulatory network (Anderson et al., 2004; Asselbergh et al., 2008; Li et al., 2011; Atkinson and Urwin, 2012). The expression profiles of auxin-responsive genes observed in rice compatible and incompatible interactions with *Magnaporthe grisea* and *Striga hermonthica* demonstrate a role of auxin in plant defense responses and suggest crosstalk between auxin abiotic and biotic stress pathways. At least 154 auxin-induced and 61 auxin-repressed genes were found to be differentially expressed under one or more of these biotic stress conditions. Among the 50 members of auxin-related gene families (*GH*3, *Aux/IAA*, *SAUR* and *ARF*), 14 showed responses to both *M. grisea* and *S. hermonthica* infection, with the remaining groups of 20 and 16 genes exhibiting specific responses to *M. grisea* and *S. hermonthica* infection, respectively. Overlapping responses under both biotic and abiotic stress conditions revealed that several auxin-responsive genes and members of auxin-related gene families responded to both biotic and abiotic stresses (Ghanashyam and Jain, 2009; Jain and Khurana, 2009).

CONCLUSION

The balance and interplay of hormone systems play pivotal roles in the expression of pathogenesis and resistance during plant-pest interactions. The effectiveness of these mechanisms relies in large part on perception of pathogen-derived molecules by plant receptors, which triggers the biosynthesis of phytohormones. New insights into plant defense signaling network indicate that auxin participates in several stress responses. Notably, a series of downstream events in the auxin-signaling pathway are activated and may promote susceptibility or resistance to both biotrophic or necrotrophic pathogens, depending on pathogen lifestyles and the host genetic constitution. Furthermore, there is evidence that auxin expression also mediates adaptive responses to abiotic stress. In this complex scenario, Arabidopsis defective mutants continue to encompass an important tool for elucidating the auxin-related processes that mediate stress/defense responses. Despite several recent advances, the exact mechanism of auxin-mediated stress/defense response remains unclear, and further investigations are needed to unveil the direct and indirect molecular mechanisms that modulate plant-pathogen interactions.

REFERENCES

Anderson, J.P., Badruzsaufari, E., Schenk, P.M. & et al. (2004). Antagonistic interaction between abscisic acid and jasmonate-ethylene signaling pathways modulates defence gene expression and disease resistance in Arabidopsis. *Plant Cell*, 16:3460–3479.

Asselbergh, B., Achuo, A.E., Hofte, M. &Van Gijsegem, F. (2008). Abscisic acid deficiency leads to rapid activation of tomato defence responses upon infection with *Erwinia chrysanthemi*. *Mol Plant Pathol.*, 9:11–24.

Atkinson, N.J. & Urwin, P.E. (2012). The interaction of plant biotic and abiotic stresses: from genes to the field. *J Exp Bot.*, 63(10):3523-43.

Azevedo, C., Betsuyaku, S., Peart, J., Takahashi, A., Noe¨, L., Sadanandom, A., Casais, C., Parker, J. & Shirasu, K. (2006). Role of SGT1 in resistance protein accumulation in plant immunity. *Embo J.*, 25, 2007–2016.

Bari, R. & Jones, J.D.G. (2009). Role of plant hormones in plant defence responses. *Plant Mol. Biol.*, 69, 473-488.

Beneventi, M.A., da Silva Jr, O.B., Sá, M.E.L., Firmino, A.A.P., de Amorim, R.M.S., Albuquerque, E.V.S., Silva, M.C.M., Silva, J.P., Campos, M.A., Lopes, M.J.C., Togawa, R.C., Pappas Jr, G.J. & Grossi–de–Sa, M.F. (2013). Transcription profile of soybean-root-knot nematode interaction reveals a key role of phythormones in the resistance reaction. *BMC Genomics*, 14:322.

Benjamins, R. & Scheres, B. (2008). Auxin: the looping star in plant development. *Annu. Rev. Plant Biol.*, 59:443–465.

Berrocal-Lobo, M. & Molina, A. (2004). Ethylene response factor 1 mediates Arabidopsis resistance to the soilborne fungus *Fusarium oxysporum*. *Mol. Plant-Microbe Interact.*, 17, 763–770.

Chen, Z., Agnew, J.L., Cohen, J.D., He, P., Shan, L. & et al. (2007). *Pseudomonas syringae* type III effector AvrRpt2 alters *Arabidopsis thaliana* auxin physiology. *Proc. Natl. Acad. Sci. USA*, 104, 20131–20136.

Chen, Z., Kloek, A.P., Cuzick, A., Moeder, W., Tang, D. & et al. (2004). The *Pseudomonas syringae* type III effector AvrRpt2 functions downstream or independently of SA to promote virulence on *Arabidopsis thaliana*. *Plant Journal*, 37, 494–504.

Dai, Y., Wang, H., Li, B., Huang, J., Liu, X., Zhou, Y., Mou, Z. & Li, J. (2006). Increased expression of MAP KINASE KINASE7 causes deficiency in polar auxin transport and leads to plant architectural abnormality in Arabidopsis. *Plant Cell*, 18, 308-320.

Ding, X., Cao, Y., Huang, L., Zhao, J., Xu, C., Li, X. & Wang, S. (2008). Activation of the indole-3-acetic acid-amido synthetase GH3-8 suppresses expansin expression and promotes salicylate- and jasmonate-independent basal immunity in rice. *Plant Cell*,20, 1, 228-240.

Durrant, W.E. & Dong, X. (2004). Systemic acquired resistance. *Annu. Rev. Phytopathol*,. 42: 185–209.

Eshraghi, L., Anderson, J.P., Aryamanesh, N., McComb, J.A., Shearer, B. & Hardy, G.St.J.E. (2014). Suppression of the auxin response pathway enhances susceptibility to *Phytophthora cinnamomi* while phosphite-mediated resistance stimulates the auxin signalling pathway. *BMC Plant Biology*, 14, 68, 1-15.

Fu, J. &Wang, S. (2011). Insights into auxin signaling in plant–pathogen interactions. *Frontiers in Plant Sci*., 2, 74, 1-7.

Ghanashyam, C. & Jain, M. (2009). Role of auxin-responsive genes in biotic stress responses. *Plant Signl. & Behav*., 4,9, 846-848.

Gimenez-Ibanez, S. & Solano, R. (2013). Nuclear jasmonate and salicylate signaling and crosstalk in defense against pathogens. *Frontiers in Plant Defense*, 4, 1-11.

Glickmann, E., Gardan, L., Jacquet, S., Hussain, S., Elasri, M., Petit, A. & Dessaux,Y. (1998). Auxin production is a common feature of most pathovars of *Pseudomonas syringae*. *Mol. Plant Microbe Interact*., 11,156-162.

Hanashyam, C. & Jain, M. (2009). Role of auxin-responsive genes in biotic stress responses. *Plant Signal. & Beh*., 4:9, 846-848.

Jain, M. & Khurana, J.P. (2009). Transcript profiling reveals diverse roles of auxin-responsive genes during reproductive development and abiotic stress in rice. *FEBS J*., 276, 3148-3162.

Jones, J.D. & Dangl, J.L. (2006). The plant immune system. *Nature*, 444:323–329.

Kazan, K. & Manners, J.M. (2009). Linking development to defense: auxin in plant-pathogen interactions. *Trends Plant Sci*., 7, 373–382.

Korasick, D.A., Enders, T.A., Strader L.C. (2013). Auxin biosynthesis and storage forms. *J. Exp. Bot*., 64: 2541–2555.

Llorente, F., Muskett, P., Sánchez-Vallet, A., López, G., Ramos, B., Sánchez-Rodrıguez, C., Jordá, L., Parker, J. & Molina, A. (2008). Repression of the auxin response pathway increases Arabidopsis susceptibility to necrotrophic fungi. *Mol. Plant*, 1, 3,496–509.

Manulis, S., Haviv-Chesner, A., Brandl, M.T., Lindow, S.E. & Barash, I. Differential involvement of indole-3-acetic acid biosynthetic pathways in pathogenicity and epiphytic fitness of *Erwinia herbicola* pv. *Gypsophilae*. (1998). *Mol. Plant Mic. Interac*., 11, 7, 6334-642.

Maor, R., Haskin,S., Levi-Kedmi, H. & Sharon, A. (2004). In planta production of indole-3-acetic acid by *Colletotrichum gloeosporioides* f. sp. *aeschynomene*. *Appl. Environ.Microbiol*., 70, 1852–1854.

Marois, E., Van den Ackerveken, G. & Bonas, U. (2002). The xanthomonas type III effector protein AvrBs3 modulates plant gene expression and induces cell hypertrophy in the susceptible host. *Mol. Plant-Microbe Interact*., 15, 637–646.

Mockaitis, K. & Estelle, M. (2008). Auxin receptors and plant development: a new signaling paradigm. *Annu. Rev. Cell Dev. Biol.,* 24, 55-80.

Naseem, M., Philippi, N., Hussain, A., Wangorsch, G., Ahmed, N. & et al. (2012). Integrated systems view on networking by hormones in Arabidopsis immunity reveals multiple crosstalk for cytokinin. *Plant Cell*, 24, 1793–1814.

Naseem, M. & Dandekar, T. (2012). The role of auxin-cytokinin antagonism in plant-pathogen interactions. *PLoS Pathog*., 8, 11, 1-4.

Navarro, L., Dunoyer, P., Jay F., Arnold, B., Dharmasiri, N., Estelle, M. & et al. A plant miRNA contributes to antibacterial resistance by repressing auxin signaling. (2006). *Science*, 312,436-439.

Park, C. (2007). Auxin homeostasis in plant stress adaptation response. *Plant Signl. & Behav*., 224, 306-307.

Park, J.E., Park, J.Y., Kim, Y.S., Staswick, P.E., Jeon, J., Yun, J., Kim, S.Y., Kim, J., Lee, Y.H. & Park, C.M. (2007). GH3-mediated auxin homeostasis links growth regulation with stress adaptation response in Arabidopsis. *J. Biol. Chem*., 282, 13, 10036-10046.

Pieterse, C.M., Leon-Reyes, A., Van der Ent, S. & Van Wees, S.C. (2009). Networking by small-molecule hormones in plant immunity. *Nat. Chem. Biol*., 5, 5, 308–316.

Quint, M., Barkawi, L. S., Fan, K., Cohen, J.D. & Gray, W.M. (2009). Arabidopsis IAR4 modulates auxin response by regulating auxin homeostasis. *Plant Physiol*., 150, 748-758.

Robert-Seilaniantz, A., Grant, M. & Jones, J.D.G. (2011). Hormone crosstalk in plant disease and defense: more than just jasmonate-salicylate antagonism. *Annu. Rev. Phytopathol.*, 49, 317–343.

Robert-Seilaniantz, A., Navarro, L., Bari, R., Jones, J.D. (2007). Pathological Santner, A. & Estelle, M. (2009). Recent advances and emerging trends in plant hormone signaling. *Nature*, 459,,1071-1078.

Sá, M.E.L., Lopes, M.J.C., Campos, M.A., Paiva, L.V., Santos, R.M.A., Beneventi, M.A., Firmino, A.A.P., Grossi-de-Sa, M.F. (2012). Transcriptome analysis of resistant soybean roots infected by *Meloidogyne javanica*. *Genet. & Mol. Biol.*, 35, 1 (suppl), 272-282.

Siemens, J., Keller, I., Sarx, J., Kunz, S., Schuller, A. & et al. (2006). Transcriptome analysis of *Arabidopsis clubroots* indicate a key role for cytokinins in disease development. *Mol. Plant-Microbe Interact.*, 19, 480–494.

Spaepen, S. & Vanderleyden, J. (2011). Auxin plant-microbe interactions. *Cold Spring Harb. Perspect. Biol.*, 3, 1-13.

Vandeputte, O., Oden, S., Mol, A., Vereecke, D., Goethals, K., El Jaziri, M. & Prinsen, E. (2005). Biosynthesis of auxin by the gram-positive phytopathogen *Rhodococcus fascians* is controlled by compounds specific to infected plant tissues. *Appl. Environ. Microbiol.*, 71,3, 1169-1177.

Wang, D., Mukhtar, K.P., Culler, A.H. & Dong, X. (2007). Salicylic acid inhibits pathogen growth in plants through repression of the auxin signaling pathway. *Curr. Biol.*, 17, 1784–1790.

Westfall, C.S., Herrmann, J., Chen, Q., Wang, S. & Jez, J.M. (2010). Modulating plant hormones by enzyme action: the GH3 family of acyl acid amido synthetases. *Plant Signl.& Behav.*, 61, 752–766.

Woodward, A.W. & Bartel, B. (2005). Auxin: regulation, action and interaction. *Ann. Bot.*, 95, 707–735.

Yang, S., Zhang, Q., Guo, J., Charkowski, A. O., Glick, B.R., Ibekwe, A.M., Cooksey, D.A. &Yang, C.H. (2007). Global effect of indole-3- acetic acid biosynthesis on multiple virulence factors of *Erwinia chrysanthemi* 3937. *Appl. Environ. Microbiol.*, 73, 1079–1088.

Zhang, X., Dai, Y., Xiong, Y., DeFraia, C., Li, J., Dong, X. & Mou, Z. (2007). Overexpression of Arabidopsis MAP kinase kinase 7 leads to activation of plant basal and systemic acquired resistance. *Plant Journal*, 52, 6, 1066-1079.

Zhang, Z., Li, Q., Li, Z., Staswick, P.E., Wang, M., Zhu, Y. & et al. (2007). Dual regulation role of GH3.5 in salicylic acid and auxin signaling during

Arabidopsis-*Pseudomonas syringae* interaction. *Plant Physiol.*, 145:450-464.

Zhao, Y., Christensen, S.K., Fankhauser, C., Cashman, J.R., Cohen, J.D., Weigel, D. & Chory, D. (2001). A role for flavin monooxygenase-like enzymes in auxin biosynthesis. *Science*, 291, 5502, 306-309.

In: Auxin
Editor: Rosario D. Pacovsky
ISBN: 978-1-61122-060-5

Chapter 3

How Does Auxin Influence on Plant Developmental Responses during Salinity?

***M. J. Iglesias*[1], *M. C. Terrile*[1] *and C. A. Casalongué*[1,*]**
[1]Instituto de Investigaciones Biológicas,
UE CONICET-Universidad Nacional de Mar del Plata,
Funes 3250 (7600) Mar del Plata,
Argentina

Abstract

Plants integrate intrinsic developmental programs and quickly adjust growth to environmental stimuli. Plant growth and development largely depend on the phytohormone auxin which exerts its function throughout partially redundant F-box receptors, TIR1-AFBs. Several studies using pharmacological and genetic approaches demonstrated that auxin response is down-regulated under salinity. However, how auxin regulation impacts on growth and developmental programs is still under study. Developmental plasticity of roots is modulated by auxin and also is critical for tolerance to stress. Reasonably, plant acclimation to salinity could involve root growth reprogramming depending on the intensity of the stress whereby plant growth is redirected to diminish stress exposure. It is hypothesized that stress-induced morphological responses (SIMR)

* E-mail: casalong@mdp.edu.ar, Fax: 54 223 4753030.

are mediated by auxin-mediated redox metabolism. However, one emerging question is how salinity transduces stress signals to regulate root plasticity during plant acclimation. The entwined mechanisms involving miRNAs and Salt Overly Sensitive (SOS) pathway in the regulation of auxin homeostasis will be also discussed.

INTRODUCTION

Soil salinity is one of the most important abiotic stresses that affect agricultural productivity worldwide. Salinity triggers several metabolic, cellular and physiological responses leading to acclimation in plant growth (Golldack et al., 2011; Shabala and Munns, 2012).

In the case of mild stress, plant acclimation involves the development of SIMR. Potters et al. (2007) postulated such induced responses as part of a plant acclimation strategy whereby growth and development is reprogrammed to minimize exposure to stress. Therefore, the study of the molecular mechanisms that mediate SIMR promotion is essential to understand plant responses to salt stress. Since plants are sessile, root plasticity is a relevant mechanism in order to survive under changeable environments. For example, in Arabidopsis mild salt stress results in a drastic reduction of primary root (PR) and lateral root (LR) elongation but not necessarily in the number of LR (Zolla et al., 2010). On the other hand, high concentrations of salt, over 100 mM NaCl, trigger inhibition of LR development (Burssens et al., 2000; Deak and Malamy, 2005; He et al., 2005). Therefore, depending on the intensity and duration of stress and the plant developmental stage, root architecture could be differentially modulated (Potters et al., 2009; Wang et al., 2009).

In this revision, we briefly summarize some understood aspects of auxin-mediated regulation in particular, focused on roots. Then, the analysis of auxin and redox crosstalk together with the modulation of auxin homeostasis by miRNA and SOS signaling will be particularly addressed.

MODULATION OF AUXIN REGULATION UNDER SALINITY

Auxin is a phytohormone with critical roles in almost every aspect of growth and development as well as in responses to stress along the plant life-cycle. It is critical in controlling PR elongation and LR development including primordium formation, LR initiation, emergence and elongation (Casimiro et al., 2003). In this context, modulation of auxin homeostasis plays a significant function in acclimation to stress (Park et al., 2007; Wolters and Jurgens, 2009). To correctly direct functions, auxin must be distributed, located and perceived in specific cells and tissues. In consequence, its homeostasis is under different regulation levels to ensure an optimized control of biosynthesis, conjugation-hydrolysis, degradation, transport and perception.

The most important and abundant auxin in plants is the indole-3-acetic acid (IAA). In general, it is synthesized through two major pathways: the tryptophan (Trp)-dependent and the Trp-independent route (Normanly, 2010). These IAA biosynthetic pathways provide many points of additional regulation. Higher plants store IAA in the form of conjugates; IAA can be ester-linked to sugars or amide-linked to amino acids or peptides. The functions of these conjugates include storage, transport, compartmentalization and stability (Woodward and Bartel, 2005). Specifically, GH3s are auxin-inducible genes that encode auxin-conjugating enzymes (Staswick et al., 2005). IAA-conjugate hydrolases release free IAA from the respective conjugates (Bartel and Fink, 1995). These enzymes regulate endogenous free IAA pool. One of the first studies that proposed the regulation of auxin conjugation during stress responses was described by Park et al (2007). Arabidopsis *wes1-D* mutant, in which a *GH3* gene is overproduced, shows growth reduction evidenced by a dwarf phenotype, reduction in the LR number and enhanced tolerance to various stresses including salinity. In rice, Zhang et al. (2009) demonstrated that activation of *GH3* gene, *TDL1/OsGH3.13*, in *tld1-D* mutant results in IAA deficiency and dramatic changes in plant architecture with enhanced tolerance to drought, suggesting that stress-induced turnover of free auxin by the GH3 enzymes could be likely part of the adaptive response to salinity.

Auxin is distributed from the tissues where is synthesized into the sink tissues by passive and active mechanisms. The passive, long-range mechanism is throughout the phloem; meanwhile, the short-range mechanism involves influx and efflux carriers. The family of AUXIN RESISTANT 1/LIKE AUX1 (AUX/LAX) genes corresponds to auxin influx carriers (Bennett et al., 1996). Among efflux carriers, PIN-FORMED (PIN) and MULTIDRUGE-RESISTANTE/P-GLICOPROTEIN (MDR/PGP) of the ATP-binding cassette family B were described (Tanaka et al., 2006; Blakeslee et al., 2007; Mravec et al., 2008; Titapiwatanakun et al., 2009).

Interestingly, high salinity induces endogenous accumulation and redistribution of auxin which leads to root architecture reprogramming by inhibition of PR growth, reduction in LR primordial formation and increment in the elongation of existing LRs (Wang et al., 2009). During mild salt exposition, restructuration of root architecture is also observed, with a reduction in PR and LR length and proliferation of LRs which are typical traits of SIMR. Since in the transporter mutant *aux1-7* the proliferation of LR primordial is completely abolished under salinity, it is suggested that changes on auxin distribution in the Arabidopsis root could promote acclimation to stress (Zolla et al., 2010).

In addition to transport, other major levels of hormonal regulation are perception and signaling. Once hormone is correctly spatio-temporary localized, it needs a correct perception to transduce the signal. In the case of auxin, the nuclear signaling pathway involves $SCF^{TIR/AFB}$-Aux/IAA co-receptor complex (Calderon Villalobos et al., 2012). TIR1 and AFB proteins specifically bind auxin (Dharmasiri et al., 2005; Dharmasiri et al., 2005; Kepinski and Leyser, 2005). Both of them are components of the SCF-type E3 ubiquitin ligase that target the Aux/IAA repressor proteins for the degradation by the 26S proteasome which in turn, de-represses the transcriptional auxin response (Gray et al., 2001). The double mutant in the auxin receptors of TIR1/AFB family, *tir1 afb2* shows tolerance to salinity suggesting that down-regulation of TIR1/AFBs-mediated signaling could also be part of Arabidopsis acclimation (Iglesias et al., 2010).

Next, we tried to summarize the mechanisms that could contribute to the modulation of auxin homeostasis in salt-stressed roots.

AUXIN HOMEOSTASIS AND SOS SIGNALING

Since sodium accumulation is toxic for several metabolic reactions, one of the responses to salt stress is the maintainance of cellular ion homeostasis through the activation of the SOS signaling pathway which consists of three main components SOS3, SOS2, and SOS1 (Zhu, 2000). Sodium perception by an unknown sensor induces cytosolic calcium signaling mediating SOS1 phosphorylation by the SOS3 /SOS2 kinase complex (Halfter et al., 2000; Liu et al., 2000). SOS1, encoding a Na^+/H^+ antiporter is responsible for sodium extrusion (Quintero et al., 2002).

In the last years, evidence on SOS signaling revealed its crosstalk with other signaling pathways. In particular, the crosstalk between SOS and auxin signaling pathways has been reported in the regulation of root development under salt stress (Ji et al., 2013). In response to non-uniform salinity, root growth is dynamically redirected to avoid salinity. This avoidance tropism is mediated by altering auxin distribution in roots with the concomitant repression of gravitropic responses (Li and Zhang, 2008). Sun et al. (2008) described that both, auxin accumulation and tropism are impaired under salinity in *sos2* mutant, revealing a SOS regulation of auxin homeostasis. Alternatively, in a uniform saline environment a plant strategy to reduce exposure to stress includes a reduction in root hair number and length. Root hair development is an auxin-mediated response (Rahman et al., 2002). Arabidopsis *sos* mutants exhibited defective phenotypes in root hair under salinity supporting the fact that an interlink between auxin and SOS might occur in response to salt (Wang et al., 2008).

In addition, Zhao et al. (2011) demonstrated that SOS signaling is essential for activation of LR primordial and emergence of LR under mild stress. Auxin accumulation on *sos3-1* roots was markedly reduced affecting cell specification of pericycle founder cells and cell cycle progression in newly formed LR primordial. SOS3 regulates auxin distribution through the modulation of auxin biosynthesis, polar transport and auxin gradient formation and maintenance in stressed roots. In conclusion, SOS pathway role in root plasticity involves downstream modulation of auxin homeostasis (Ji et al., 2013). In addition, Kamei et al. (2005) demonstrated that different auxin responsive genes were altered in *sos* mutants. IAA5 induction in *sos2-1* and *sos3-1* suggests that SOS pathway could also negatively regulate auxin signaling during salt stress.

Auxin Homeostasis Mediated by miRNAs

MicroRNAs (miRNA) and short interfering RNAs (siRNAs) are abundant classes of tiny and single-stranded noncoding RNAs. In multicellular organisms including plants, these molecules have emerged with important roles in the regulation of protein-coding, translational repression, and chromatin modification (Ambros et al., 2003). It is well-accepted that abiotic stresses regulate the expression of thousands of genes at both, transcriptional and posttranscriptional levels. In this direction, post-transcriptional gene regulation might have a critical role in the plant response to salt stress. Furthermore, root development is critically regulated by auxin but also by miRNAs (Meng et al., 2010).

One of the first global studies on the identification of miRNAs in the tolerance to abiotic stress was performed through library sequencing in Arabidopsis plants. In response to cold, dehydration, NaCl and ABA treatments, miR393 was strongly up-regulated in Arabidopsis (Sunkar and Zhu, 2004). Several miRNAs, including miR393 were also up-regulated by salt treatment using microarray data (Liu et al., 2008). Interestingly, members of the family of auxin receptors TIR1/AFBs were described as targets of miR393 suggesting that auxin signaling might be repressed under salt at posttranscriptional level (Navarro et al., 2006).

In Arabidopsis, miR393 is encoded by two loci, *AtMIR393A* and *AtMIR393B* producing different precursors but an identical mature miRNA (Jones-Rhoades and Bartel, 2004). Si-Amomour et al. (2011) demonstrated that MIR393B and secondary siRNAs regulate TIR1/AFB2 gene expression and auxin-related development in Arabidopsis leaves. In rice, auxin receptor gene orthologs *OsTIR1* and *OsAFB2* and the auxin transporter *AUX1* are targets of OsmiR393 resulting in an additional level of complexity of auxin signaling homeostasis. Two members of the miR393 family were found in rice, *Os-MIR393A* and *Os-MIR393B*. However, specifically *Os-MIR393A* gene expression changes by salinity and alkaline stress (Gao et al., 2011). The overexpression of Arabidopsis *AtMIR393A* gene in tobacco modified auxin response and enhanced tolerance to salt stress (Feng et al., 2010). *AtMiIR393A* has been also involved in TIR1 posttranscriptional repression during biotic stress, suggesting that a locus-specific induction of *AtMIR393A* expression during stress could involve an additional layer of regulation of auxin-signaling pathway (Navarro et al., 2006). Endorsing more evidence about the role of miR393 in the tolerance to stress, an stress-regulated miR393-guided cleavage of TIR1 and AFB2 transcripts is required for inhibition of LR growth by ABA and osmotic stress, suggesting that the attenuation of auxin signaling modulates root adaptation to salinity (Chen et al., 2012).

In maize, the miRNA genotype-specific expression might explain the distinct salt sensitivities between salt-tolerant and salt–sensitive lines. Even more, known stress-responsive elements AREs (anaerobic induction elements) and two ABREs (ABA-response elements) were found in the promoters of different members of miR393 family in maize, providing a possible evidence for ABA-mediated regulation of auxin signaling repression in response to salt (Ding et al., 2009).

In leaves, miR160-directed Auxin Response Factor 17, ARF17 regulation is important for proper development and expression of early auxin response genes (Mallory et al., 2005). Meanwhile, in roots the negative regulation of ARF10 by miR160 was demonstrated to play a role in early developmental programs by mechanisms involving interactions between ARF10-dependent auxin and ABA pathways (Liu et al., 2007), suggesting that miRNA–mediated regulation could be critical for early auxin response.

In a more recent study, Kinoshita et al. (2012) described that miR167-IAR3 interaction plays a role in root architecture under high osmotic and drought stress. The amidohydrolase enzyme, IAR3 releases free auxin from an inactive precursor.

All these evidences highlighted varied functions of miRNAs in the regulation of auxin at posttranscriptional level (Mangrauthia et al., 2013). However, the field of miRNA biology is rather incipient in the plant response to stress.

AUXIN AND REDOX INTERLINK DURING ROOT DEVELOPMENT UNDER SALINITY

The accumulation of Reactive Oxygen Species (ROS) is a common response to abiotic stress including salinity (Mittler, 2002). Major ROS include superoxide anion (O_2^{-}), hydroxyl radical (·OH), singlet oxygen ($^1O_2$) and hydrogen peroxide (H_2O_2). In the last years, the redox biology has witnessed a dramatic reappraisal on the importance of ROS. It is well- known that ROS have a dual function since, besides being products of aerobic metabolism, they are also signaling molecules involved in multiple biological processes, including programmed cell death, stomatal movement, growth and development events (Foyer and Shigeoka, 2011).

Both, ROS and auxin have been suggested as key elements in the adjustment of growth and developmental programs during acclimation to stress (Foyer and Noctor, 2012; Tognetti et al., 2012). The relationships between auxin and redox signalings are a paramount in the root architecture adjustment (De Tullio et al., 2010). Wang et al. (2010) described that exogenous H_2O_2 treatments mimic auxin-mediated LR induction. Auxin-mediated ROS production was also reported in adventitious root formation in mung bean (Bai et al., 2012). Much more recently, auxin-mediated ROS accumulation (H_2O_2 y $O_2^{\cdot -}$) and peroxidase activities were demonstrated to be required during LR formation and elongation (Ma et al., 2013; Manzano et al., 2014).

Likely, auxin mediated ROS production involves the modulation of the NAD(P)H oxidase RbohD activity (Peer et al., 2013) and the repression of components of the antioxidant system. Auxin negatively regulates the expression and activity of ascorbate peroxidase 1 (APX1) in Arabidopsis (Iglesias et al., 2010; Correa-Aragunde et al., 2013) while induces ascorbate oxidase activity which catalyzes the oxidation of ascorbate to DHA in tobacco plants (Pignocchi et al., 2003). Ascorbate redox state appears as a regulation point during root development since an auxin-mediated increment in the oxidized form of dehydroascorbate (DHA) in roots of tomato (Solanum lycopersicum) leads to PR growth arrest and LR elongation (Tyburski et al., 2008).

This is consistent with alterations in PR elongation and LR formation in response to auxin described in the *apx1* mutant (Correa-Aragunde et al., 2013). Therefore, since auxin response is down-regulated under salinity, ROS metabolism and redox-dependent physiological processes could be affected during stress. In this context, a genetic approach by using the auxin mutants, *tir1 afb2* revealed an auxin-ROS interaction during salinity (Iglesias et al., 2010). The *tir1 afb2* plants show a reduced accumulation of H_2O_2 y $O_2^{\cdot-}$, higher APX enzymatic activity as well as increased levels of ascorbate and changes in root.

In summary, down-regulation of auxin response may constitute a regulatory module by which plants redirect root architecture through the modulation of redox metabolism during acclimation to salinity. However, the study of the stress signals and underlying mechanisms involved in the auxin and redox regulation under salinity is still in its beginning.

Conclusion and Future Research Perspectives

In conclusion, salt stress effects can be observed at physiological, biochemical and molecular levels in plants. However, root becomes the first organ committed to the stress because of reduction in water ingress and then, by exposition to ionic toxicity. This work describes essential roles for auxin in the acclimation of roots to salinity. An integrative modulation of auxin homeostasis may be a regulatory module involved in SIMR leading to root plasticity throughout the interlink with ROS-associated metabolism (Figure 1).

However, despite the advances achieved on the knowledge of tolerance to salinity, important issues such as, compartmentalization and spatial and temporal relationships between auxin and ROS signaling should be considered and studied in the near future.

Additionally, miRNAs could have a significant role in orchestrating gene expression, and functions to modulate growth and root development under salt stress. The possibility to develop salt stress tolerance in crop plants using miRNAs could be also a near challenge.

The SOS signaling pathway could also function beyond its role in Na^+ extrusion in the regulation of auxin homeostasis during the reprogramming of root development under salinity.

Taking together all these findings it is possible to speculate that modulation of auxin signaling by salinity could be regulated by the SOS pathway, different miRNAs and other hormonal crosstalk-mediated events. Since miRNAs not only regulate the expression of genes but also modulate its own functions it is likely that a more complex scenario can take part in the hormonal regulation under salt stress.

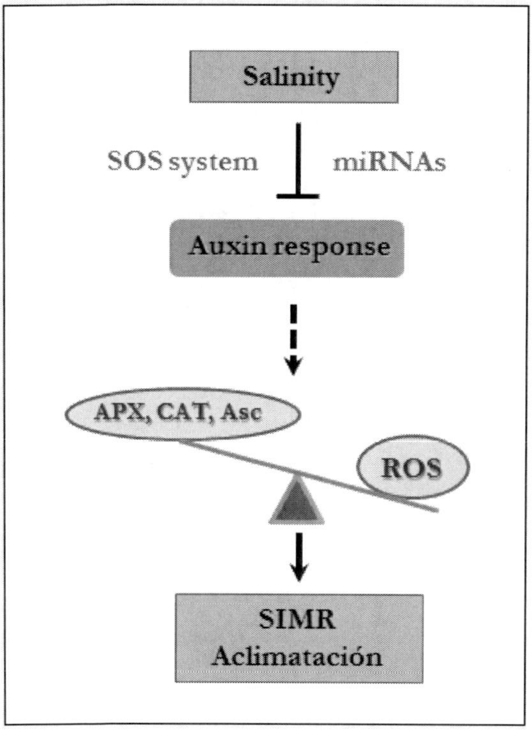

Figure 1. Hypothetical model for auxin regulation mediated by SOS and miRNAs in the stress induced morphological response, SIMR to salt stress. SOS pathway and different miRNAs are proposed as sensing molecules involved in the regulation of auxin homeostasis during root acclimation. Auxin modulation might directly or indirectly regulates the redox status involved in root plasticity and SIMR. Repression of auxin homeostasis might contribute to a more efficient distribution of resources to develop tolerance to stress.

ACKNOWLEDGMENT

This work was partially supported by grants from ANPCyT, CONICET and UNMDP. MCT and CAC are members of the research staff of CONICET. MJI is a doctoral fellow of the UNMDP. We thank Dr Diego Fiol for his critical reading of our manuscript.

REFERENCES

Ambros V., Bartel B., Bartel D., Burge C., Carrington J., Chen X., Dreyfuss G., Eddy, Griffiths-Jones S., Marshall M., Matzke M., Ruvkun G., Tuschl T. (2003) A uniform system for microRNA annotation. *RNA*, 9: 277-279.

Bai M.-Y., Fan M., Oh E., Wang Z.-Y. (2012) A Triple Helix-Loop-Helix/Basic Helix-Loop-Helix Cascade Controls Cell Elongation Downstream of Multiple Hormonal and Environmental Signaling Pathways in Arabidopsis. *The Plant Cell*, 24: 4917-4929.

Bartel B., Fink G. R. (1995) ILR1, an amidohydrolase that releases active indole-3-acetic acid from conjugates. *Science,* 268: 1745-1748.

Bennett M. J., Marchant A., Green H. G., May S. T., Ward S. P., Millner P. A., Walker A. R., Schulz B., Feldmann K. A. (1996) Arabidopsis AUX1 gene: a permease-like regulator of root gravitropism. *Science,* 273: 948-950.

Blakeslee J. J., Bandyopadhyay A., Lee O. R., Mravec J., Titapiwatanakun B., Sauer M., Makam S. N., Cheng Y., Bouchard R., Adamec J., Geisler M., Nagashima A., Sakai T., Martinoia E., Friml J., Peer W. A., Murphy A. S. (2007) Interactions among PIN-FORMED and P-glycoprotein auxin transporters in Arabidopsis. *The Plant Cell*, 19: 131-147.

Burssens S., Himanen K., van de Cotte B., Beeckman T., Van Montagu M., Inzé D., Verbruggen N. (2000) Expression of cell cycle regulatory genes and morphological alterations in response to salt stress in Arabidopsis thaliana. *Planta*, 211: 632-640.

Calderon Villalobos L. I., Lee S., De Oliveira C., Ivetac A., Brandt W., Armitage L., Sheard L. B., Tan X., Parry G., Mao H., Zheng N., Napier R., Kepinski S., Estelle M. (2012) A combinatorial TIR1/AFB-Aux/IAA co-receptor system for differential sensing of auxin. *Nature Chemical Biology,* 8: 477-485.

Casimiro I., Beeckman T., Graham N., Bhalerao R., Zhang H., Casero P., Sandberg G., Bennett M. J. (2003) Dissecting Arabidopsis lateral root development. *Trends in Plant Science*, 8: 165-171.

Correa-Aragunde N., Foresi N., Delledonne M., Lamattina L. (2013) Auxin induces redox regulation of ascorbate peroxidase activity by S-nitrosylation/denitrosylation balance resulting in changes of root growth pattern in Arabidopsis. *Journal of Experimental Botany*, 64: 3339-3349.

Chen H., Li Z., Xiong L. (2012) A plant microRNA regulates the adaptation of roots to drought stress. *FEBS Letters*, 586: 1742-1747.

De Tullio M. C., Jiang K., Feldman L. J. (2010) Redox regulation of root apical meristem organization: Connecting root development to its environment. *Plant Physiology and Biochemistry*, 48: 328-336.

Deak K. I., Malamy J. (2005) Osmotic regulation of root system architecture. *The Plant Journal,* 43: 17-28.

Dharmasiri N., Dharmasiri S., Estelle M. (2005) The F-box protein TIR1 is an auxin receptor. *Nature,* 435: 441-445.

Dharmasiri N., Dharmasiri S., Weijers D., Lechner E., Yamada M., Hobbie L., Ehrismann J. S., Jurgens G., Estelle M. (2005) Plant development is regulated by a family of auxin receptor F box proteins. *Developmental Cell,* 9: 109-119.

Ding D., Zhang L., Wang H., Liu Z., Zhang Z., Zheng Y. (2009) Differential expression of miRNAs in response to salt stress in maize roots. *Annals of Botany,* 103: 29-38.

Feng X.-M., You C.-X., Qiao Y., Mao K., Hao Y.-J. (2010) Ectopic overexpression of Arabidopsis AtmiR393a gene changes auxin sensitivity and enhances salt resistance in tobacco. *Acta Physiologia Plantarum*, 32: 997-1003.

Foyer C. H., Noctor G. (2012) Managing the cellular redox hub in photosynthetic organisms. *Plant Cell Environment*, 35: 199-201.

Foyer C. H., Shigeoka S. (2011) Understanding oxidative stress and antioxidant functions to enhance photosynthesis. *Plant Physiology*, 155: 93-100.

Gao P., Bai X., Yang L., Lv D., Pan X., Li Y., Cai H., Ji W., Chen Q., Zhu Y. (2011) osa-MIR393: a salinity- and alkaline stress-related microRNA gene. *Molecular Biology Reports*, 38: 237-242.

Golldack D., Luking I., Yang O. (2011) Plant tolerance to drought and salinity: stress regulating transcription factors and their functional significance in the cellular transcriptional network. *Plant Cell Reports*, 30: 1383-1391.

Gray W. M., Kepinski S., Rouse D., Leyser O., Estelle M. (2001) Auxin regulates SCF(TIR1)-dependent degradation of AUX/IAA proteins. *Nature,* 414: 271-276.

Halfter U., Ishitani M., Zhu J. K. (2000) The Arabidopsis SOS2 protein kinase physically interacts with and is activated by the calcium-binding protein SOS3. *Proccedings of the National Academy of Sciences U S A*, 97: 3735-3740.

He X.-J., Mu R.-L., Cao W.-H., Zhang Z.-G., Zhang J.-S., Chen S.-Y. (2005) AtNAC2, a transcription factor downstream of ethylene and auxin signaling pathways, is involved in salt stress response and lateral root development. *The Plant Journal*, 44: 903-916.

Iglesias M., Terrile M., Bartoli C., D'Ippólito S., Casalongué C. (2010) Auxin signaling participates in the adaptative response against oxidative stress and salinity by interacting with redox metabolism in Arabidopsis. *Plant Molecular Biology,* 74: 215-222.

Ji H., Pardo J. M., Batelli G., Van Oosten M. J., Bressan R. A., Li X. (2013) The Salt Overly Sensitive (SOS) pathway: established and emerging roles. *Molecular Plant*, 6: 275-286.

Jones-Rhoades M. W., Bartel D. P. (2004) Computational identification of plant microRNAs and their targets, including a stress-induced miRNA. *Molecular Cell*, 14: 787-799.

Kamei A., Seki M., Umezawa T., Ishida J., Satou M., Akiyama K., Zhu J.-K., Shinozaki K. (2005) Analysis of gene expression profiles in Arabidopsis salt overly sensitive mutants sos2−1 and sos3−1. *Plant, Cell and Environment*, 28: 1267-1275.

Kepinski S., Leyser O. (2005) The Arabidopsis F-box protein TIR1 is an auxin receptor. *Nature*, 435: 446-451.

Kinoshita N., Wang H., Kasahara H., Liu J., MacPherson C., Machida Y., Kamiya Y., Hannah M. A., Chua N.-H. (2012) IAA-Ala Resistant3, an Evolutionarily Conserved Target of miR167, Mediates Arabidopsis Root Architecture Changes during High Osmotic Stress. *The Plant Cell,* 24: 3590-3602.

Li X., Zhang W. (2008) Salt-avoidance tropism in Arabidopsis thaliana. *Plant Signaling and Behaviour*, 3: 351-353.

Liu H.-H., Tian X., Li Y.-J., Wu C.-A., Zheng C.-C. (2008) Microarray-based analysis of stress-regulated microRNAs in Arabidopsis thaliana. *RNA,* 14: 836-843.

Liu J., Ishitani M., Halfter U., Kim C. S., Zhu J. K. (2000) The Arabidopsis thaliana SOS2 gene encodes a protein kinase that is required for salt tolerance. *Proccedings of the National Academy of Sciences U S A,* 97: 3730-3734.

Ma F., Wang L., Li J., Samma M., Xie Y., Wang R., Wang J., Zhang J., Shen W. (2013) Interaction between HY1 and H2O2 in auxin-induced lateral root formation in Arabidopsis. *Plant Molecular Biology*, 1-13.

Mallory A. C., Bartel D. P., Bartel B. (2005) MicroRNA-directed regulation of Arabidopsis AUXIN RESPONSE FACTOR17 is essential for proper development and modulates expression of early auxin response genes. *The Plant Cell,* 17: 1360-1375.

Mangrauthia S. K., Agarwal S., Sailaja B., Madhav M. S., Voleti S. R. (2013) MicroRNAs and their role in salt stres response in plants. *In* P Ahmad, MM Azooz, MNV Prasad, eds, Salt stress in Plants. Signallings, omics and adaptations. Springer Heildelber, New York.

Manzano C., Pallero M., Casimiro I., De Rybel B., Orman-Ligeza B., Van Isterdael G., Beeckman T., Draye X., Casero P., del Pozo J. C. (2014) The emerging role of ROS signalling during lateral root development. *Plant Physiology*.

Meng Y., Ma X., Chen D., Wu P., Chen M. (2010) MicroRNA-mediated signaling involved in plant root development. *Biochemical and Biophysical Reseach Communications,* 393: 345–349.

Mittler R. (2002) Oxidative stress, antioxidants and stress tolerance. *Trends in Plant Science,* 7: 405-410.

Mravec J., Kubes M., Bielach A., Gaykova V., Petrasek J., Skupa P., Chand S., Benkova E., Zazimalova E., Friml J. (2008) Interaction of PIN and PGP transport mechanisms in auxin distribution-dependent development. *Development,* 135: 3345-3354.

Navarro L., Dunoyer P., Jay F., Arnold B., Dharmasiri N., Estelle M., Voinnet O., Jones J. D. (2006) A plant miRNA contributes to antibacterial resistance by repressing auxin signaling. *Science,* 312: 436-439.

Normanly J. (2010) Approaching cellular and molecular resolution of auxin biosynthesis and metabolism. *Cold Spring Harbor Perspective Biology*, 2: a001594.

Park J. E., Park J. Y., Kim Y. S., Staswick P. E., Jeon J., Yun J., Kim S. Y., Kim J., Lee Y. H., Park C. M. (2007) GH3-mediated auxin homeostasis links growth regulation with stress adaptation response in Arabidopsis. *Journal of Biological Chemistry*, 282: 10036-10046.

Peer W. A., Cheng Y., Murphy A. S. (2013) Evidence of oxidative attenuation of auxin signalling. *Journal of Experimental Botany*, 64: 2629-2639.

Pignocchi C., Fletcher J. M., Wilkinson J. E., Barnes J. D., Foyer C. H. (2003) The Function of Ascorbate Oxidase in Tobacco. *Plant Physiology*, 132: 1631-1641.

Potters G., Pasternak T. P., Guisez Y., Jansen M. A. K. (2009) Different stresses, similar morphogenic responses: Integrating a plethora of pathways. *Plant, Cell and Environment*, 32: 158-169.

Potters G., Pasternak T. P., Guisez Y., Palme K. J., Jansen M. A. (2007) Stress-induced morphogenic responses: growing out of trouble? *Trends in Plant Science,* 12: 98-105.

Quintero F. J., Ohta M., Shi H., Zhu J.-K., Pardo J. M. (2002) Reconstitution in yeast of the Arabidopsis SOS signaling pathway for Na+ homeostasis. *Proceedings of the National Academy of Sciences U S A*, 99: 9061-9066.

Rahman A., Hosokawa S., Oono Y., Amakawa T., Goto N., Tsurumi S. (2002) Auxin and ethylene response interactions during Arabidopsis root hair development dissected by auxin influx modulators. *Plant Physiology*, 130: 1908-1917.

Shabala S., Munns R. (2012) Salinity stress: physiological constraints and adaptative mechanism. *In* S S, ed, Plant stress physiology. CABI, pp 59-93.

Si-Ammour A., Windels D., Arn-Bouldoires E., Kutter C., Ailhas J., Meins F., Vazquez F. (2011) miR393 and Secondary siRNAs Regulate Expression of the TIR1/AFB2 Auxin Receptor Clade and Auxin-Related Development of Arabidopsis Leaves. *Plant Physiology*, 157: 683-691.

Staswick P. E., Serban B., Rowe M., Tiryaki I., Maldonado M. T., Maldonado M. C., Suza W. (2005) Characterization of an Arabidopsis enzyme family that conjugates amino acids to indole-3-acetic acid. *The Plant Cell,* 17: 616-627.

Sun F., Zhang W., Hu H., Li B., Wang Y., Zhao Y., Li K., Liu M., Li X. (2008) Salt Modulates Gravity Signaling Pathway to Regulate Growth Direction of Primary Roots in Arabidopsis. *Plant Physiology*, 146: 178-188.

Sunkar R., Zhu J. K. (2004) Novel and stress-regulated microRNAs and other small RNAs from Arabidopsis. *The Plant Cell*, 16: 2001-2019.

Tanaka H., Dhonukshe P., Brewer P. B., Friml J. (2006) Spatiotemporal asymmetric auxin distribution: a means to coordinate plant development. *Cell Molecular Life Sciences*, 63: 2738-2754.

Titapiwatanakun B., Blakeslee J. J., Bandyopadhyay A., Yang H., Mravec J., Sauer M., Cheng Y., Adamec J., Nagashima A., Geisler M., Sakai T., Friml J., Peer W. A., Murphy A. S. (2009) ABCB19/PGP19 stabilises

PIN1 in membrane microdomains in Arabidopsis. *The Plant Journal*, 57: 27-44.

Tognetti V. B., Muhlenbock P., Van Breusegem F. (2012) Stress homeostasis - the redox and auxin perspective. *Plant, Cell and Environment*, 35: 321-333.

Tyburski J., Krzemiński Ł., Tretyn A. (2008) Exogenous auxin affects ascorbate metabolism in roots of tomato seedlings. *Plant Growth Regulation,* 54: 203-215.

Wang P., Du Y., Li Y., Ren D., Song C.-P. (2010) Hydrogen Peroxide–Mediated Activation of MAP Kinase 6 Modulates Nitric Oxide Biosynthesis and Signal Transduction in Arabidopsis. *The Plant Cell*, 22: 2981-2998.

Wang Y., Li K., Li X. (2009) Auxin redistribution modulates plastic development of root system architecture under salt stress in Arabidopsis thaliana. *Journal of Plant Physiology*, 166: 1637-1645.

Wang Y., Zhang W., Li K., Sun F., Han C., Wang Y., Li X. (2008) Salt-induced plasticity of root hair development is caused by ion disequilibrium in *Arabidopsis thaliana. Journal of Plant Research*, 121: 87-96.

Wolters H., Jurgens G. (2009) Survival of the flexible: hormonal growth control and adaptation in plant development. *Nature Review Genetics*, 10: 305-317.

Woodward A. W., Bartel B. (2005) Auxin: regulation, action, and interaction. *Annals in Botany (Lond.)*, 95: 707-735.

Zhang S. W., Li C. H., Cao J., Zhang Y. C., Zhang S. Q., Xia Y. F., Sun D. Y., Sun Y. (2009) Altered architecture and enhanced drought tolerance in rice via the down-regulation of indole-3-acetic acid by TLD1/OsGH3.13 activation. *Plant Physiology*, 151: 1889-1901.

Zhao Y., Wang T., Zhang W., Li X. (2011) SOS3 mediates lateral root development under low salt stress through regulation of auxin redistribution and maxima in Arabidopsis. *New Phytologist*, 189: 1122-1134.

Zhu J.-K. (2000) Genetic Analysis of Plant Salt Tolerance Using Arabidopsis. *Plant Physiology*, 124: 941-948.

Zolla G., Heimer Y. M., Barak S. (2010) Mild salinity stimulates a stress-induced morphogenic response in Arabidopsis thaliana roots. *Journal of Experimental Botany,* 61: 211-224.

In: Auxin
Editor: Rosario D. Pacovsky
ISBN: 978-1-61122-060-5

Chapter 4

HISTORY, SYNTHESIS, TRANSPORT, DETERMINATION, METABOLISM, PROPERTIES AND PHYSIOLOGICAL EFFECTS OF AUXINS IN PLANT GROWTH

Virginia Sarropoulou[*], *Kortessa Dimassi-Theriou and Ioannis Therios*
Laboratory of Pomology, Department of Horticulture,
School of Agriculture, Aristotle University of Thessaloniki,
Thessaloniki, Greece

ABSTRACT

The word auxin has a Greek origin: *auxein* means to enlarge or to grow. Auxins are an important category of plant hormones, regulating many aspects of plant growth and development throughout the plant cell cycle, from cell division, cell elongation and vascular differentiation to root initiation, somatic embryogenesis, apical dominance, tropistic responses, flowering, fruit ripening, senescence, parthenocarpy, dormacy and pollen maturation. Common purification procedures for the determination of auxins include column chromatography, solid phase extraction, liquid-liquid and mass spectrometric detection. The four naturally occurring (endogenous) auxins are indole-3-acetic acid (IAA),

[*] Corresponding author: Email: vsarrop@gmail.com; vsarropo@agro.auth.gr.

4-chloroindole-3-acetic acid, phenylacetic acid and indole-3-butyric acid (IBA); only these four were found to be synthesized by plants. Synthetic auxin analogs include 1-naphthaleneacetic acid (α-NAA), 2,4-dichlorophenoxyacetic acid (2,4-D) and many others. The most important member of the auxin family is IAA. Most of the IAA produced within plants is conjugated to other compounds to form esters, amides or glycosyl esters. The most commonly occurring IAA-conjugates are indole-3-acetylaspartic acid (IAAsp) and a range of IAA glucose esters (IAAGlu). Conjugation seems to be a mechanism for storing IAA in cells, stabilising the level of free auxin in the plant, and metabolising its excess. Auxin in conjugated molecules is protected from oxidative breakdown and may be released again through the action of enzymes. Natural and synthetic auxins are low-molecular weight, organic substances, containing either an indole or an aromatic ring. They are crystalline and only slightly soluble in water, but readily soluble in organic solvents (ethanol, methanol, acetone, diethyl ether and dimethyl sulphoxide, DMSO) or, as weak acids, in alkaline watery solutions. In intact plants, the highest level of IAA occurs predominantly in or near the apex and IAA is transported basipetally; thus, IAA concentration decreases from the top to the bottom. The above is especially true for monocotyledons, whereas in dicotyledons the highest concentration of IAA occurs in the subapical zone, which also grows most rapidly. Auxin is also abundant in young leaves, floral organs and developing fruits and seeds. IAA is photo-oxidized rapidly (50 % in 24 h) while IBA is oxidized slowly (10 %) and NAA is very stable. IAA concentrations are high in young, fast growing organs and decline with age. The indole amino acid, tryptophan (Trp) is the precursor of IAA. There are several possibilities as to how plants may convert Trp into IAA: via indole-3-pyruvic acid, via indole-3-acetaldoxime and *via* tryptamine. Auxins exhibit active transport in a polar manner in addition to long-distance movement via vascular tissues.

1. History of Auxins and Pioneering Experiments

Auxins were the first plant hormones discovered. Charles Darwin was among the first scientists to dabble in plant hormone research. In his book "The Power of Movement in Plants" presented in 1880, he first describes the effects of light on movement of canary grass (*Phalaris canariensis*) coleoptiles. The coleoptile is a specialized leaf originating from the first node which sheaths the epicotyl in the plants seedling stage protecting it until it emerges from the ground. When unidirectional light shines on the coleoptile, it bends in the direction of the light. If the tip of the coleoptile was covered with aluminum foil, no bending would occur towards the unidirectional light.

However if the tip of the coleoptile was left uncovered but the portion just below the tip was covered, exposure to unidirectional light resulted in curvature toward the light. Darwin's experiment suggested that the tip of the coleoptile was the tissue responsible for perceiving the light and producing some signal which was transported to the lower part of the coleoptile where the physiological response of bending occurred. He then cut off the tip of the coleoptile and exposed the rest of the coleoptile to unidirectional light to see if curving occurred. Curvature did not occur confirming the results of his first experiment [1].

It was in 1885 that Salkowski discovered indole-3-acetic acid (IAA) in fermentation media [2]. The isolation of the same product from plant tissues would not be found in plant tissues for almost 50 years. IAA is the major auxin involved in many of the physiological processes in plants [3]. In 1907, Fitting studied the effect of making incisions on either the light or dark side of the plant. His results were aimed at understanding if translocation of the signal occurred on a particular side of the plant but his results were inconclusive because the signal was capable of crossing or going around the incision [4]. In 1913, Boysen-Jensen modified Fritting's experiment by inserting pieces of mica to block the transport of the signal and showed that transport of auxin toward the base occurs on the dark side of the stem as opposed to the side exposed to the unidirectional light [5]. In 1918, Paal confirmed Boysen-Jensen's results by cutting off coleoptile tips in the dark, exposing only the tips to the light, replacing the coleoptile tips on the plant but off centered to one side or the other. Results showed that whichever side was exposed to the coleoptile, curvature occurred toward the other side [6]. Soding was the next scientist to extend auxin research by extending on Paal's idea. He showed that if tips were cut off there was a reduction in growth but if they were cut off and then replaced growth continued to occur [7].

In 1926, a graduate student from Holland by the name Fritz Went published a report describing how he isolated a plant growth substance by placing agar blocks under coleoptile tips for a period of time then removing them and placing them on decapitated Avena stems [8]. After placement of the agar, the stems resumed growth. In 1928, Went developed a method of quantifying this plant growth substance. His results suggested that the curvatures of stems were proportional to the amount of growth substance in the agar [9]. This test was called the *avena curvature test.* Much of our current knowledge of auxin was obtained from its applications. Went's work had a great influence in stimulating plant growth substance research. He is often credited with dubbing the term auxin but it was actually Kogl and Haagen-

Smit who purified the compound auxentriolic acid (auxin A) from human urine in 1931 [10]. Later Kogl isolated other compounds from urine which were similar in structure and function to auxin A, one of which was indole-3 acetic acid (IAA) initially discovered by Salkowski in 1985. In 1954 a committee of plant physiologists was set up to characterize the group auxins. The term comes from the Greek auxein meaning "to grow." Compounds are generally considered auxins if they are synthesized by the plant and are substances which share similar activity to IAA (the first auxin to be isolated from plants) [11, 3].

2. Roles of Auxins General

Auxin also plays a major role in lateral root initiation and development. Lateral root development can be divided in different steps: primordium initiation and development, emergence, and meristem activation. Furthermore, it is also involved in the growth and organization of lateral root primordia and emergence from the parent root [12]. However, the overall auxin content is not always positively correlated with lateral root number [13]. Auxin transport into the regions where lateral root initiate also seems crucial for the regulation of root branching [14]. In the root, auxin is transported in both an acropetal (base-to-apex) [15] and basipetal (apex-to-base) direction within inner and outer tissues of the root apex [16]. Studies using auxin transport inhibitors have shown that the polarity of auxin movement provides an important developmental signal during lateral root initiation [14] and root gravitropism [17].

The role of auxins in morphogenesis is evident from the fact that the cells which respond to auxin revert to a dedifferentiated state and begin to divide. It was reported that auxins cause DNA to become more methylated than usual and they suggested that this might be necessary for the re-programming of differentiated cells [18].

3. Auxin Homeostasis in the Vascular Cambium

Several substances with auxin activity have been identified in plants. Indole-3-acetic acid (IAA) is both the most abundant endogenous auxin, and the most widely studied [11]. Recently, it was shown that all parts of young

Arabidopsis plants have the ability to produce IAA [15], but it is believed that the major sites of IAA biosynthesis in older plants are apical shoots and young leaves [11]. IAA is transported through plant tissues in a polar fashion, but can also be transported via the phloem [20, 21]. In the case of tree stems, the main pathway for polar IAA transport is the cambial zone and its most recent derivatives [22]. The polar transport is slow, about 5-20 mm h-1 [23], and mediated by specific influx and efflux carriers located in the plasma membrane of the cells [24]. Auxin is important in controlling the regulation of many different cellular and developmental processes in plants. This control is mediated not only through changes in the cellular auxin concentration, but also through changes in tissue sensitivity towards auxin and changes in the polar auxin transport [11, 24, 25]. IAA is distributed in a steep radial gradient across the cambial region tissues, with the highest concentration in the cambial zone and decreasing concentrations towards the maturing xylem and the phloem [22]. Since the rate of polar transport is slow and the distance from the apical shoot to the basal part of the stem can be considerable in a large tree, additional IAA sources are likely to contribute to maintaining the IAA pool in the cambium. Both *de novo* synthesis of IAA in the cambium and addition of IAA from mature leaves have been proposed as additional sources of the IAA present in the cambial region [26]. Another putative source of IAA is hydrolysis of auxin stored as amide- and ester conjugates. Most plant organs contain significant pools of hydrolysable IAA conjugates, but in wood-forming tissues of softwoods and hardwoods only trace amounts have been found [27, 28]. The concentration gradient of IAA across the cambial region tissues may be maintained by lateral diffusion followed by removal of IAA when reaching the xylem and phloem transport systems, where endogenous IAA has been found [29].

4. Auxin and Wood Formation

It is well established that IAA is an important regulator of cambial growth [30, 31, 22]. In decapitated stems, apically applied IAA enters the polar transport system and mimics the IAA supply from the apical shoot by inducing cambial cell division and xylem production in a dose-dependent manner [32, 33]. Exogenous IAA also affects the radial diameter of xylem elements [34] and induces tracheary differentiation in callus and cell suspension cultures [35].

5. Determination of Auxins

Auxins are difficult to analyze because they occur in very low amounts in culture extracts, which are quite rich in interfering substances [36, 37]. Common purification procedures including column chromatography, solid phase extraction, and liquid-liquid extraction are used to purify auxin and indole compounds [38]. However, these procedures generally require significant amounts of solvents, time, and labor [39, 38].

Mass spectrometric detection is a well-recognized and specific method that enables the sensitive and accurate measurement of auxin and indole compounds in crude extracts [40, 38].

6. Discovery of Auxins

The discovery of auxins during the nineteenth century was the outcome of experiments on phototropism and geotropism [41]. In 1880, Charles Darwin reported on the phenomenon by which the plants bent toward the sunlight, in a book entitled "*The power of movement of plants*". Several years later, by 1926, the Dutch botanist Fritz W. Went discovered auxin and described a bioassay for its quantitative detection by "the *Avena* coleoptile curvature test".

Although Went had succeeded in isolating auxin, he was not able to purify the active compound to establish its chemical structure. In 1934, the biochemists Kögl, Haagen-Smit and Erxleben obtained an active substance from urine, indole-3-acetic acid (IAA), which was found to be identical to auxin. Finally, K.V. Thimann isolated IAA from cultures of the fungus *Rhizopus suinus* in 1935.

The first generally accepted report of the occurrence of IAA in a higher plant was published by Haagen-Smit et al., in 1946 [41]. Since then, there have been an increasing number of reports of the occurrence in plants of IAA and other related compounds, such as indole-3-butyric acid (IBA), 4-chloro-IAA, and conjugated IAA forms, and non-indolic compounds, such as phenylacetic acid that displays weak auxin activity [42].

7. Effect and Role of Plant Hormones on the Plant Physiology and Development

IAA represents one of the most important plant hormones, regulating many aspects of plant growth and development throughout the plant cell cycle, from cell division, cell elongation and differentiation to root initiation, apical dominance, tropistic responses, flowering, fruit ripening and senescence. Regulation of these processes by auxin is believed to involve auxin-induced changes in gene expression [43]. There is no complete description of the mechanism by which auxin regulates cell growth. The immediate effect of exposure of plants tissues to auxin is proton excretion, occurring within minutes. The resulting apoplastic acidification provides a favourable condition for cell wall loosening, which could be an early part of auxin-induced cell expansion [44].

Auxins are very widely used in plant tissue culture and usually form an integral part of nutrient media. Auxins promote, mainly in combination with cytokinins, the growth of calli, cell suspensions and organs, and also regulate the direction of morphogenesis. The word auxin has a Greek origin: *auxein* means to enlarge or to grow. At the cellular level, auxins control basic processes such as cell division and cell elongation. Since they are capable of initiating cell division they are involved in the formation of meristems giving rise to either unorganised tissue, or defined organs. In organised tissues, auxins are involved in the establishment and maintenance of polarity and in whole plants their most marked effect is the maintenance of apical dominance and mediation of tropisms [45].

The choice of auxins and the concentration administered depends on:

- the type of growth and/or development required
- the rate of uptake and transport of the applied auxin to the target tissue;
- the inactivation (oxidation and/or conjugation) of auxin in the medium and within the explant
- the natural levels and the endogenous synthesis within the explant
- the sensitivity of the plant tissue to auxin (and other hormones as well)
- the interaction, if any, between applied auxins and the natural endogenous substances.

The four naturally occurring (endogenous) auxins are IAA, 4-chloroindole-3-acetic acid, phenylacetic acid and indole-3-butyric acid; only these four were found to be synthesized by plants. Synthetic auxin analogs include 1-naphthaleneacetic acid, 2,4-dichlorophenoxyacetic acid (2,4-D) and many others. The most important member of the auxin family is indole-3-acetic acid (IAA) [46]. On the molecular level, all auxins are compounds with an aromatic ring and a carboxylic acid group [47].

In 1926, the Dutch botanist Fritz Warmolt Went showed that a chemical messenger diffuses from coleoptile tips. Went's experiment identified how a growth promoting chemical causes a coleoptile to grow towards light. Went cut the tips of the coleoptiles and placed them in the dark, putting a few tips on agar blocks that he predicted would absorb the growth-promoting chemical. On control coleoptiles, he placed a block that lacked the chemical. On others, he placed blocks containing the chemical, either centred on top of the coleoptile to distribute the chemical evenly or offset to increase the concentration on one side [48]. When the growth promoting chemical was distributed evenly the coleoptile grew straight. If the chemical was distributed unevenly, the coleoptile curved away from the side with the cube, as if growing towards light, even though it was grown in the dark. Went later proposed that the messenger substance is a growth-promoting hormone, which he named auxin, that becomes asymmetrically distributed in the bending region. Went concluded that auxin is at a higher concentration on the shaded side, promoting cell elongation, which results in a coleoptiles bending towards the light (Mendipweb Nature of auxin).

8. Naturally Occurring and Synthetic Auxins

The most commonly detected natural auxin is IAA (indole-3-acetic acid); but endogenous occurrence of 4-chloro-IAA (2) [49] and of indole-3-butyric acid (IBA) [50] have also been demonstrated. Furthermore, the weak auxin phenylacetic acid (PAA) occurs naturally in plants [51] and there are precursors and metabolites of IAA present in plant tissues, like indole-3-pyruvic acid, tryptamine [52] or tryptophol [53, 54]. In addition, the intermediate of agrobacterial IAA biosynthesis, indole-3-acetamide, has been detected in plant tissues [55].

Most of the IAA produced within plants is conjugated to other compounds to form esters, amides or glycosyl esters. The most commonly occurring IAA-conjugates are indole-3-acetylaspartic acid (IAAsp) and a range of IAA glucose esters (IAAGlu). Conjugation seems to be a mechanism for storing IAA in cells, stabilising the level of free auxin in the plant, and metabolising its excess [19]. Auxin in conjugated molecules is protected from oxidative breakdown and may be released again through the action of enzymes.

IAA may be used as an auxin in plant tissue culture media, but it tends to be oxidised in culture media and is rapidly metabolised within plant tissues. However, this characteristic can be useful, because in some plants, callus induced by IAA (together with cytokinins) frequently gives rise to shoots or embryos as its effective concentration becomes diminished. IAA has also been used with other regulants to induce direct morphogenesis (including the rooting of microcuttings), and for meristem and shoot cultures [56]. Also, in this case, decreasing IAA concentration may be important: in apple microcuttings high IAA at the beginning of culture induces root formation but, later on, the growth of roots is promoted by lower IAA concentrations [57]. However, for many purposes, it is necessary or desirable to use one of the many synthetic analogues of IAA. These analogues have different structures but similar biological properties and are also called auxins. The synthetic auxins 2,4-dichlorophenoxyacetic acid (2,4-D) and naphthalene acetic acid (NAA) are not oxidised, but they can also be converted to conjugates with glucose.

9. Physical and Chemical Properties of Auxins

Natural and synthetic auxins are low-molecular weight, organic substances, containing either an indole or an aromatic ring. They are crystalline and only slightly soluble in water, but readily soluble in organic solvents (ethanol, methanol, acetone, diethyl ether and dimethyl sulphoxide, DMSO) or, as weak acids, in alkaline aqueous solutions. With the exception of IAA they are stable and persist in the media for tissue cultures. IAA is less stable and is especially sensitive to the light (mainly UV) and to oxidants. Both naturally occurring indolic auxins and their synthetic counterparts, in spite of their different structures, have similar physiological effects. For auxin activity, an aromatic ring is needed at a distance of 0.55 Å from a carboxyl group. Indeed, all active auxins are weak organic acids. The relative degree of activity of individual auxins in different growth processes is very variable. It differs not only from plant to plant, but also from organ to organ, tissue to tissue, cell to cell and, moreover, also with the age and physiological state of the plant (tissue) [58]. Probably due to its high instability, IAA is usually less effective than synthetic auxins like 2,4-D or NAA.

Auxin is a weak organic acid consisting of a planar indole ring structure coupled to a side chain harbouring a terminal carboxyl group. The carboxyl group is protonated at low pH, making the molecule less polar (IAA– + H+ ⇔ IAA-H). In this form it can diffuse across cell membranes, whereas the molecule in its unprotonated, negatively charged form (IAA–) is too polar to diffuse. The pH in different cellular compartments varies, being ~5.0-5.5 in the apoplastic fluid of the cell wall and in vacuoles and ~7.0 in the cytosol. IAA-H in the apoplast and in vacuoles can thus diffuse over the surrounding membranes, whereas IAA– is trapped within the cell and cannot escape from the cytosol without the aid of specific transporters (efflux carriers) [59, 60].

10. Metabolism of Auxin

10.1. Natural Levels of IAA

In intact plants, the highest level of IAA occurs predominantly in or near the apex and IAA is transported basipetally; thus, we can observe a decreasing IAA concentration from the top to the bottom. The above is especially true for monocotyledons, whereas in dicotyledons the highest concentration of IAA occurs in the subapical zone, which also grows most rapidly [61]. Auxin is also abundant in young leaves, floral organs and developing fruits and seeds. IAA concentrations are high in young, fast growing organs and decline with age and are affected by external factors, e.g. by light (wavelength, intensity and photoperiod). The level is not constant during the day: daily oscillations in IAA concentration in *Chenopodium rubrum* plants have been described. There are no significant changes in constant light, but an endogenous rhythm has been observed in constant darkness. Various diurnal fluctuations were observed in different photoperiodic regimes [62]. Levels of naturally occurring auxin in explanted tissues are found to depend on the mother plant from which the explants are taken. The age of the mother plant, the conditions under which it has been growing and the season of the year at which explants are taken, can all be influential [63]. As stated above, the time of the day may also be important. The presence of an associated organised meristem can cause callus to grow more vigorously, or to be capable of organogenesis [64]. This suggests that meristematic cells are particularly active sites for the biosynthesis and/or the release of natural growth factors favouring cell growth [65].

10.2. Auxin Biosynthesis

Traditionally, the indole amino acid, tryptophan (Trp), which is formed as well as the other aromatic compounds by the shikimic acid pathway, has been thought to be a precursor of the most important native auxin in plants, IAA.

There are several possibilities as to how plants may convert Trp into IAA: *via* indole-3-pyruvic acid (the so called "indolepyruvate pathway"), *via* indole-3-acetaldoxime (the "indole-3-acetaldoxime pathway") and *via* tryptamine (the "tryptamine pathway"). The most common IAA-biosynthetic pathway in plants appears to be the indolepyruvate one; it begins by the transamination of Trp catalysed by the enzyme tryptophan transaminase (or tryptophan amino transferase) [66]. Indolepyruvate is then converted into indole-3-acetaldehyde [67]. IAA can originate from this aldehyde either by dehydrogenation (NAD-dependent indoleacetaldehyde dehydrogenase) or by oxidation (indoleacetaldehyde oxidase) [68].

The indole-3-acetaldoxime pathway starts by formation of indole-3-acetaldoxime followed by its conversion to indole-3-acetonitrile and final hydrolysis (nitrilase) to IAA [69]. This pathway is typical for, *e.g.*, Brassicaceae and Resedaceae. The tryptamine pathway was reported mainly in Poaceae and involves decarboxylation of Trp into tryptamine (tryptophan decarboxylase) followed by its conversion to indole-3-acetaldehyde. At this point this pathway joins with the indolepyruvate pathway. Since one of the enzymes converting indole-3-acetaldehyde into IAA, *i.e.*, indoleacetaldehyde oxidase, shows sigmoidal kinetics, its regulatory role for both these pathways has been proposed [70]. There is one more IAA biosynthetic pathway starting from Trp: it takes place in some plant pathogenic bacteria and consequently also in plants transformed by *Agrobacterium*. The first step in this pathway is the formation of indole-3-acetamide (Trp monooxygenase) and its subsequent hydrolysis to IAA (indole-3-acetamide hydrolase) [71].

IAA biosynthesis in plants can occur via different pathways [72], which are classified according to their intermediates: indole-3-acetamide (IAM), indole-3-pyruvate (IPyA), tryptamine, and indole-3-acetonitrile [73]. The two most common routes for IAA biosynthesis in bacteria are the IAM and the IPyA pathways. Two major pathways for IAA biosynthesis have been proposed in plants: the tryptophan (Trp)-independent and (Trp)-dependent pathways [74, 75, 76]. In Trp-independent IAA biosynthesis, indole-3-glycerol phosphate or indole is the likely precursor, but little is known about the biochemical pathway to IAA In Trp-dependent IAA biosynthesis, several pathways have been postulated: (i) the indole-3-acetamide (IAM) pathway; (ii) the indole-3-pyruvic acid (IPA) pathway; (iii) the tryptamine (TAM) pathway; and (iv) the indole-3-acetaldoxime (IAOX) pathway [74, 75, 76, 77].

Synthesis of auxin involves two pathways with tryptophan dependent and tryptophan independent pathway. Auxin can be inactivated thorough conjugation and oxidation [74].

10.3. Auxin Conjugation and Degradation

Auxin is known to form two main classes of conjugates: amides (peptides) and glycosyl esters. Amides originate *via* formation of peptide bonds between auxins (both native and synthetic) and amino acids, namely aspartate, glutamate, alanine, glycine and valine. Glycosyl esters of IAA involve compounds such as various isomers of *o*-indole-3-acetyl-β-D-glucopyranose and indole-3-acetyl-*myo*inositol, indole-3-acetylglucosyl-rhamnose, etc. [19].

In contrast to conjugation reactions, auxin degradation is an irreversible process. There are two main ways of IAA degradation: oxidative decarboxylation and oxidation without decarboxylation. The oxidative decarboxylation may be catalysed by a non-specific activity traditionally called "IAA oxidase/peroxidase" and leads to the subsequent formation of degradation products such as indole-3-methanol, indole-3-aldehyde, methylene oxindole and indole-3-carboxylic acid [78].

Oxidation without decarboxylation involves complex pathways resulting in the formation of oxindole-3-acetic and dioxindole-3-acetic acids and their derivatives. Similar pathways are functional also for degradation of some conjugates, *e.g.*, indole-3-aspartate [78].

The synthetic auxins, 2,4-D and NAA are often converted, after uptake into plant tissues, to conjugates, mainly glucosyl esters [79].

10.4. Effect of Synthetic Auxins on IAA Levels

The synthetic auxins, which are added exogenously to control the growth and organisation of cultured tissues, may affect endogenous IAA levels. This can be caused by inhibition of IAA oxidase.

Reducing the external 2,4-D and NAA concentration resulted in a significant increase in internal free IAA concentration in the auxin-dependent and cytokinin autonomous tobacco cell strain VBI-0 [80]. In carrot hypocotyl explants, neither 2,4-D nor NAA, both of which induced callus formation, had any effect on endogenous IAA concentration. This shows that in this case synthetic auxins induced morphogenesis themselves [81].

10.5. Stability in Culture Media

IAA and to some extent also IBA are heat labile and decompose during autoclaving. IAA is also unstable in culture media at room temperature. In the dark, there can be more than a ten-fold decrease in concentration over a four-week period in the absence of inocula [82]. The rate of decrease of IAA is even more rapid in the light and is accelerated by the presence of MS salts [83]. In liquid MS medium incubated at 25°C in a normal 16 h photoperiod, 10 μM (1.75 mg/l) IAA was reduced to less than the limit of detection (0.05 mg/l) in 14 days [82]. Because, for oxidative degradation, oxygen is required, in solid media the degradation of IAA is likely to be significantly less. IBA is more stable in solution than IAA: 75% remained unaltered after 30 days in the dark, and 40% after 30 days in the light. Other auxins such as NAA and 2,4-D are not oxidized [83], but once absorbed within plant tissues, the rate of either degradation or conjugation of all auxins may be rapid, for they are then not only exposed to physical factors, but are also subjected to enzymatic conversion. It has been shown that in tissue culture media, IAA is photo-oxidized rapidly (50% in 24 h) while IBA oxidized slowly (10%) and NAA is very stable [84].

11. Transport of Auxins

Auxins seem to be the only group of plant hormones exhibiting - on the level of the whole plant or its parts - active transport in a polar manner in addition to long-distance movement via vascular tissues [85].

The phenomenon of polar auxin transport was demonstrated first in the late 1920s [9]. In contrast to the movement of auxin *via* vascular tissues, polar cell-to-cell auxin transport is localised predominantly in parenchyma cells in the sheath surrounding differentiated vascular tissue. The polarity of auxin transport was explained [86] by the different permeability of opposite parts of cells for dissociated and undissociated molecules of indole-3-acetic acid (IAA− and IAAH, respectively), and by an asymmetric localisation (basal in the stem cells) of a so-called auxin efflux carrier (translocating the IAA anion outside the cell).

There is some evidence that cytoskeleton (microfilaments in particular) and/or Golgi-mediated protein traffic may be involved in establishment and maintenance of this distribution [87].

Auxin is thought to be synthesized in plants locally, in young growing regions, predominantly in the shoot apex, young leaves and developing seeds [88], but it seems that almost any plant tissue can at certain times be responsive for auxin [11]. Already in the 1920s, Cholodny and Went were independently trying to hypothesize how auxin moves from the apex into the elongation zone [89].

Two main pathways describe the transport of auxin, a fast, non-directional transport in the phloem and a slower, directional, so-called polar auxin transport (PAT) in various tissues. The evidence for the existence of phloem transport was established through experiments with radioactively labelled auxin [90]. This transport occurs in both basipetal and acropetal directions, proceeds relatively fast (5–20 cm/h) and seems to correlate well with the transport of assimilates and inactive auxin conjugates [91]. Direct auxin analysis revealed physiological relevant amounts of free indole-3-acetic acids (IAA) within the phloem exudates [20]. Experiments in pea showed that the labelled auxin transported within the phloem was later detected in the PAT system indicating that both transport pathways may be linked [92].

In contrast to phloem transport, PAT is specific for active free auxins, occurs in a cell-to-cell manner and has a strictly unidirectional character. The main PAT stream runs from the apex basipetally with a velocity of 5–20 mm/h towards the base of the plant [93]. Using radioactively labelled auxin this kind of transport was mainly detected in the cambium and adjacent, partially differentiated xylem elements [90]. In roots the auxin stream continues acropetally towards the root tip, where part of the auxin is redirected backwards and transported basipetally through the root epidermis to the elongation zone [94]. In contrast, PAT in shoots occurs in lateral direction [90]. Auxin transport assays also revealed that PAT requires energy, and is saturable and sensitive to protein synthesis inhibitors. These results suggested the existence of a specific auxin transport proteins and led in the middle of the 1970s to the formulation of a coherent model for auxin transport, termed the chemiosmotic hypothesis [86]. In the relatively acidic environment of the cell wall (pH around 5.5) about 15% of IAA exists in its protonated form (IAAH).

This non charged, lipophilic molecule passes easily through the plasma membrane by diffusion. In the more basic cytoplasm (pH around 7) IAAH dissociates and hence the resulting IAA−anion is 'trapped' in the cell due to its poor membrane permeability. Therefore, a specific efflux carrier was postulated and the polarity of the auxin flux was explained by its asymmetric distribution in cells. In addition, the existence of specific auxin influx carriers was hypothesized [95] and later saturable auxin influx, probably working as an IAA−/2H+ co-transporter, was physiologically demonstrated [96]. Auxin influx and efflux pathways can be physiologically distinguished using auxin efflux inhibitors (AEI). These pharmacological tools arose from correlative exploration of structure-activity profiles of chemicals with auxin-like activity [97] and were demonstrated to inhibit efflux of auxin from cells and hypocotyl segments [98]. Most widely recognized AEI such as 1-*N*-naphthylphthalamic acid (NPA) share a benzoic acid ortho-linked with an aromatic ring system. However other very effective inhibitors such as 2,3,5-triiodobenzoic acid (TIBA) do not share this structural pattern. In order to explain the mechanism of inhibition of auxin efflux by AEI, the existence of an NPA-binding protein forming part of the auxin efflux carrier complex was postulated. Additional experiments with auxin efflux and NPA binding studies in the presence of inhibitors of protein synthesis suggested the existence of a third unstable component of the auxin efflux carrier complex likely coupling the NPA-binding protein to auxin transport protein [99]. Attempts to localize NPA-binding protein within the cell through NPA binding studies led to controversial results favouring either association with cytoskeleton [100] or integral membrane localization [101]. Auxin transport is also required for initiation and elongation of lateral roots as judged by chemical, physical and genetic methods to block auxin flow [14].

12. Physiological Effects of Auxins

A stimulatory effect on cell elongation can be demonstrated in segments of coleoptiles or stems at physiologically significant IAA concentrations. Auxin is in most cases active in the concentration range 0.1–10 μM. Activation of cell elongation by auxin is mediated by increased proton efflux ('acid growth theory') [102] and/or changes in gene expression.

Auxins stimulate differentiation of vascular bundles and, as already discussed, they take part in differentiation of buds and roots [103]. Auxin is gradually canalised by a positive feedback mechanism where increasing conductivity of auxin conducting cells leads to canals of cells efficiently transporting auxin [104]. Polar transport of auxin is fundamental for the establishment and maintenance of polarity of the plant and its organs. Inhibition of polar auxin transport leads to many abnormalities and in embryos it can lead to death [105].

Auxins are known for their ability to promote adventitious root formation. This action is definitely also coupled with stimulation of cell division – increased expression of cyclin B1 and cdc2 was observed well before the first cell division [106]. IBA is by far the most commonly used auxin to obtain root initiation in conventional cuttings. It has been shown that IBA is readily converted to IAA, but it probably also has an effect on its own [107].

Polar transport of auxin is the decisive force of apical dominance [108]. Removal of the tip, the main auxin source, or inhibition of auxin transport leads to the outgrowth of axillary buds. Also dominance of fruits is mediated by auxin transport. Uneven distribution of auxin is considered to cause differential growth rates in different sides (upper/lower or irradiated/shaded) of coleoptile or root and their bending in gravitropic or phototropic reactions [45].

13. Auxin Effects in Tissue Culture

An auxin is generally required for the induction of callus from explants. Applied auxins seem to be capable of fundamentally altering the genetically programmed physiology of whole plant tissues, which had previously determined their differentiated state. Cells, which respond to auxin, revert to a dedifferentiated state and begin to divide. Lo Schiavo et al., [18] found that auxins cause DNA to become more methylated than usual and suggested that this might be necessary for the re-programming of differentiated cells. Thus, tissue-specific programmes specifically associated with differentiation would be eradicated by hypermethylation, with perhaps a small fraction of the cells reaching an ultimate state of dedifferentiation in which they become capable of morphogenesis, or embryogenesis [109].

13.1. Somatic Embryogenesis

The process of somatic embryogenesis is often initiated in media containing high levels of auxins (especially 2,4-D), but embryos usually do not develop further until the auxin concentration is reduced. Sharp et al., [110] proposed that auxin induces an embryogenic determination in a proportion of the cells in callus or suspension cultures but at the same time causes these induced cells to cease further development into embryos. It was suggested that division of the pro-embryogenic cells and their development into embryos is only resumed at lower auxin concentrations. There are, however, many recorded exceptions to this general observation, where somatic embryos are induced even in cultures grown on media devoid of auxins. It is possible that in these instances, embryogenesis has been induced by endogenous auxin, the concentration of which has then been subsequently reduced by metabolism to permit embryo formation. In alfalfa, low 2,4-D concentration gives rise to callus from leaf explants, while higher 2,4-D levels induce formation of embryo-like structures [111]. Withdrawal of auxin from the inducing medium is associated with cell death and extracellular acidification in cultures of Norway spruce [112].

Mechanism controlling pattern formation of somatic embryogenesis is very poorly understood in plants. Polar auxin transport inhibitors are important tools in assessing the role of polar auxin transport in plant development [113, 114]. The possible role of polar auxin transport in somatic embryogenesis was tested by Schiavone and Cooke [115], who treated different stages of carrot somatic embryos with TIBA and a different auxin transport inhibitor, N-(l-naphthyl) phthalamic acid (NPA). Both auxin transport inhibitors at a concentration of 1 pM are able to block the ability of somatic embryos to undergo morphogenetic transitions to the subsequent stages: globular embryos undergo persistent spherical expansion, oblong embryos (an intermediate stage in somatic embryogenesis) continue axis elongation without any cotyledon initiation, and heart embryos develop additional growth axes on their hypocotyls. Polar auxin transport inhibitors disrupt auxin efflux from the cell but their mode of action is unknown [116].

Induction of somatic embryogenesis was achieved in presence of the auxins such as 2,4-D, NAA [117], IAA [118], dicamba [119], 2,4,5-trichlorophenoxypropionic acid, indolepropionic acid, a-naphthoxy acetic acid [119] and picloram [120, 119]. Baker et al. [121] made comparisons of the level of NAA during induction, nitrogen formulation of the medium, and photoperiod on somatic embryogenesis from immature zygotic cotyledons. Over 70% embryogenesis was obtained regardless of NAA level used. However, percent embryogenesis and number of embryos were markedly lower in explants induced on NAA compared to 2,4-D. Hazra et al. [122] stated that there was no embryo production when NAA was used with embryo axes explants. Only smooth nodular outgrowths or neomorphic protuberances were obtained when 5-30 mg/l NAA was used with immature cotyledons [123]. In contrast, Sellars et al. [124] noticed high embryo production from immature embryos with constant 2 mg/l NAA. McKently [125] found that auxin type significantly affected embryogenic response in peanut embryo axis cultures. In most of the reports, use of 2,4-D resulted in high frequency somatic embryo production in groundnut [126]. Several studies have revealed that proper polar transport of auxin is a prerequisite for normal morphogenesis beyond the globular stage [127].

13.2. Auxin Uptake and Metabolism in Tissue Cultures

IAA and synthetic auxins such as NAA and 2,4-D are rapidly taken up into cultured tissues from media with a pH less than 5-6. The compounds are subsequently absorbed into cells as whole molecules (via uptake carrier or diffusion), but dissociation then causes them to be retained within the cell, because the plasmalemma is impermeable to auxin anions [128].

14. General Roles of Auxins

Auxin, one of the most extensively studied hormones, regulates cell division, cell elongation, cell differentiation, and patterning [129]. In the intact plant, auxin can inhibit root elongation, promote hypocotyl and stem elongation, mediate root and shoot gravitropism, promote lateral root and adventitious root formation, promote root hair formation, and maintain apical dominance [130]. A variety of studies suggest that auxin is also involved in patterning events such as vascular tissue differentiation and phyllotaxy [129].

IAA is a phytohormone which is known to be involved in root initiation, cell division and cell enlargement. IAA producing bacteria are believed to increase root growth and root length, resulting in greater root surface area which enables the plant to access more nutrients from soil [131]. Recently, it was also found that IAA induces an increased level of protection in plants against external stress conditions [132].

Indole-3-acetic acid (IAA) is a key regulator of many aspects of plant growth and development, including cell division and elongation, differentiation, tropisms, apical dominance, senescence, abscission, and flowering [74]. IAA plays a major role on regulating plant growth. For example, it controls vascular tissue development, cell elongation, and apical dominance [133].

In cooperation with the other hormones, auxins, with indole-3-acetic acid (IAA) as the principal auxin in higher plants, stimulate fundamental processes such as cell elongation and division [134]. However, with increasing concentrations at the cellular sites of action, IAA induces a variety of growth abnormalities, including epinasty and inhibition of shoot (decreased internode elongation and leaf area) and root growth [135]. Stimulation of ethylene biosynthesis is a common response when auxins are applied to sensitive species [136] or overproduced in transgenic plants [137].

At high concentrations, auxin inhibits root elongation and stimulates cell differentiation. Simultaneously, cell division in the root meristem and the formation of adventitious and lateral roots are promoted [138]. Auxin is crucial for proper development of the root, and is involved in several physiological mechanisms, such as gravity response [139], branching [140] and root hair growth [141]. The importance of auxin in root formation was demonstrated in tissue culture experiments, in which a high auxin : cytokinin ratio promotes root formation, whereas a low auxin:cytokinin ratio results in the regeneration of shoots [142].

Auxins initiate meristematic activity, inhibit dormant bud development, initiate root development, have a role in leaf, flower, and fruit abscission, stimulate cell division, responsible for phototropism, stimulate ethylene production, and loosen cell wall binding. Auxins are known to inhibit the development of dormant buds such as auxiliary, epicormic, and lignotuberous buds. Increased auxin concentration through exogenous application or genetic manipulations has been shown to result in enhanced lateral root formation [143].

Auxin not only regulates the development of roots but some studies have suggested a role for auxin in regulating the position of formation of lateral roots [144]. Auxin has been commercially used to induce adventitious root formation in stem cuttings of many species [145]. Most commonly, one of the natural forms of auxin, IBA (indole-3-butyric acid), is used because of its greater stability than IAA (indole-3-acetic acid), even though IAA is more abundant than IBA in plants. Several studies have compared the effects of IAA and IBA in inducing adventitious roots and find that in most cases IBA induces adventitious roots to a greater extent than IAA [146].

Auxin participates in phototropism, geotropism, hydrotropism and other developmental changes. The uneven distribution of auxin, due to environmental cues, such as unidirectional light or gravity force, results in uneven plant tissue growth, and generally, auxin governs the form and shape of plant body, direction and strength of growth of all organs, and their mutual interaction [147].

The following are some of the responses that auxin is known to cause [148, 149, 150, 11].

- Stimulates cell elongation
- Stimulates cell division in the cambium and, in combination with cytokinins in tissue culture
- Stimulates differentiation of phloem and xylem
- Stimulates root initiation on stem cuttings and lateral root development in tissue culture
- Mediates the tropistic response of bending in response to gravity and light
- The auxin supply from the apical bud suppresses growth of lateral buds
- Delays leaf senescence
- Can inhibit or promote (via ethylene stimulation) leaf and fruit abscission
- Can induce fruit setting and growth in some plants
- Involved in assimilate movement toward auxin possibly by an effect on phloem transport
- Delays fruit ripening
- Promotes flowering in Bromeliads
- Stimulates growth of flower parts
- Promotes (via ethylene production) femaleness in dioecious flowers
- Stimulates the production of ethylene at high concentrations

15. Apical Dominance

Apically derived auxin in shoots is generally thought to control apical dominance either directly via entry into lateral buds with subsequent repression of outgrowth or indirectly via some other mechanism, i.e. activation of a second inhibitor messenger, auxin-cytokinin ratio, secondary growth substances, nutrient diversion, etc. [151, 152, 153, 154, 155]. The strongest evidence supporting either a direct or an indirect role for apically-derived auxin in controlling apical dominance comes from the classical Thimann-Skoog experiment with *Vicia faba* [156].

16. Tropisms

Signals controlling tropisms often result in a concentration gradient of auxin across a responding organ and this auxin gradient is responsible for redirecting growth [48, 157, 158]. Consider, for example, a seedling that has been placed on its side. Auxin accumulates to a higher level on the lower flanks of the hypocotyls and root.

According to the Cholodony–Went hypothesis [159], an auxin gradient triggers a differential growth response in which cells on one side of the organ elongate more than cells on the other side. Because the cells are held together and cannot move apart from one another, differential growth results in the formation of a bend.

The mechanism by which auxin induces differential cell elongation during organ bending is not understood, but it is thought that auxin alters the rate at which cells elongate by inducing changes in cell wall extensibility [160, 161]. Ethylene also influences organ bending by a mechanism that is not well understood but appears to involve the modulation of auxin transport [162, 163].

17. The Discovery of Auxin and Understanding Its Role in Phototropism

According to the classical Cholodny-Went model, lateral auxin transport gives rise to phototropism. However, experiments by Shen-Miller and Gordon [164] indicated that light inhibits polar auxin transport, which led them to propose that phototropism is caused by light-mediated inhibition of polar auxin transport on the illuminated flank of a seedling. Related to this model, Naqvi [165] proposed that light-induced production of abscisic acid on the illuminated flank of the seedling causes unequal polar auxin transport.

To test the idea that unequal polar auxin transport gives rise to phototropism, Gardner et al. [166] followed radiolabeled auxin that had been asymmetrically applied to coleoptiles. Although they were not able to completely exclude a role for light-mediated inhibition of polar auxin transport, their results further demonstrated that unilateral blue light induced lateral auxin transport. At the same time, Kang and Burg [167] reported that the enhancement of pea epicotyl phototropism by red light or gibberellins did not correlate with an increase in lateral auxin transport and proposed that the magnitude of the phototropic response is determined by adjusting auxin sensitivity. Altogether, these studies supported the involvement of auxin in phototropism, but the precise mechanism of how auxin transport and signaling cause phototropism remained unknown. Models explaining root gravitropism have proposed that the growth response of plants to gravity is regulated by an asymmetric distribution of IAA [87].

18. Auxin Marks the Floral Meristem Initiation Sites and Is Required for Floral Organ Development

Auxin also acts as a local morphogenic trigger of floral organ primordial initiation in four concentric whorls around the central floral meristem [168]. Organ boundaries on the other hand, have been suggested to require low auxin levels, allowing for organ boundary gene activity and growth repression [169]. Kaufmann et al. [170], suggested that auxin regulates floral organ growth and differentiation, partially downstream of the organ identity genes. As for the majority of lateral organs, auxin also marks the apical end of the floral organ primordia [171], and determines the venation pattern. In addition to floral meristem initiation, auxin regulates other aspects of flower development including floral organ initiation, growth, and patterning, and later events that ensure reproductive success of the mature flower [172, 173].

19. Auxin Regulates both Early and Late Stages of Stamen and Pollen Development

19.1. Auxin in Early Stamen Development (Morphogenesis, Phase 1)

It has been suggested that early anther development is mainly dependent on local auxin biosynthesis [174]. The role of auxin during these early stages has not yet been determined, but it is likely that auxin participates in the differentiation, activity, and/or fate of, for example, the tapetum layer, which plays critical roles in providing enzymes for microspore release, material for pollen wall formation, and nutrients for pollen maturation [175]. It has also been suggested that the tapetum layer may supply the developing pollen grains with indole acetic acid (IAA) [176] and reduced auxin levels in the tapetum results in decreased pollen embryogenesis [177].

19.2. Auxin Is Required for Stamen Filament Elongation, Pollen Maturation, and Anther Dehiscence (Phase 2)

Auxin needs to be transported through the stamen filament not only for induction of filament elongation but also for normal pollen development [178]. However, reduced auxin biosynthesis during pollen development results in a complete loss of pollen formation, suggesting that both biosynthesis and transport is required [179].

20. Parthenocarpy

Parthenocarpy in the cucumber can be induced genetically by the incomplete dominant gene (P), or chemically by auxins, gibberellins, and auxin transport inhibitors [180, 181, 182, 183]. The action of the gene is not known.

A correlation between high auxin content and seedlessness has been shown for some cultivars of citrus and grape [184]. Auxin transport inhibitors such as 2,3, 5-triiodobenzoic acid, methyl-2-chloro-9-hydroxyflurene-9-carboxylate (chlorflurenol), and 3,3a-dihydro-2-(P-methoxyphenyl)-8H-pyrazolo-[5, 1-a]-isoindol-8-one have been shown to be the strongest inducers of parthenocarpy in cucumber.

Ovaries treated with these auxin transport inhibitors contain a higher content of compounds with auxin-like activity than non treated ovaries [185]. Elassar et al. [186] induced parthenocarpy by treating individual cucumber flowers with IAA, p-chlorophenoxyacetic acid (4-CPA), 3-naphtoxyacetic acid (B-NOA), GA_3, GA_4, and BA.

21. Role of Auxin in Fruit Development and Dehiscence

21.1. Fruit Ripening

Observations in both climacteric and non climacteric fruits, however, have revealed that decline in auxin levels and in auxin-regulated gene expression correlates with onset of ripening [187], and that ectopically applied auxin delays ripening and represses genes involved in ripening [188]. These results suggest that auxin plays an important role in regulating the onset of fruit ripening in these species, possibly by optimizing the capacity of the fruit to ripen.

The seeds are a major source of auxin, which is transported from the seeds to the rest of the fruit. It is pivotal for the development of viable seeds that seed and fruit maturation is strictly synchronized. Auxin may therefore function as a communicator to ensure such coordination.

22. Vascular Differentiation

Early physiological studies have established that the plant hormone "auxin" is important for vascular differentiation. It has long been proposed that auxin, which is polarly transported from shoot apical meristem and young leaves, induces formation of procambial cells.

Early physiological studies have clearly demonstrated that the signals for induction of procambial cell formation are derived from apex and that exogenous auxin could replace the function of apex in the induction of procambial cell formation [189, 129]. Because procambial cells are dividing cells, auxin is likely to stimulate cell proliferation by regulating the cell cycle progression [190, 191].

Auxin is distributed in a gradient across the cambial region [192, 26] indicating that different levels of auxin together with other signaling molecules such as cytokinin are important for vascular cell differentiation [189, 193, 129, 104]. Transgenic studies have shown that alterations of endogenous auxin level dramatically affect xylem formation [71, 194].

23. Dormacy

Auxin's role in endodormancy has been described in different plant species. Levels of the auxin indole-3-acetic acid (IAA) decline gradually under short day in parallel with increasing endodormancy in bay willow and silver birch [195]. The endogenous levels of IAA are low in endodormant potato tubers and increase in shoot buds prior to the onset of growth [196]. Additionally, paradormant axillary and underground adventitious buds are inhibited by auxin after growth induction by decapitation [197, 198].

24. Apical Dominance

Different terms such as "release of apical dominance", "bud initiation", "regrowth", "branching", "axillary bud outgrowth", etc. have been used to describe growth induction of lateral buds following release of paradormancy [199]. Thimann and Skoog [156] proposed that auxin is synthesized in the primary shoot apex, moves basipetally through the stem, and inhibits the growth of axillary buds. The auxin transport hypothesis emphasizes that bud outgrowth requires auxin export from lateral buds into the basipetal auxin transport stream [200]. This hypothesis suggests that if auxin transport is at capacity in the primary stem, auxin efflux from axillary buds is inhibited. Excision or blockage of auxin transport from the apical buds allows flow of auxin from axillary buds into the primary stem, activating bud outgrowth. Auxin transport capacity is higher in the primary stem of *rms* and *max* mutants compared to wild-type plants, and these mutants have increased branching [201]. Bennett et al., [201] describes that increased auxin transport capacity in stem allows simultaneous flow of auxin from the primary apex and axillary buds into the stem.

In many plants the growing shoot apex inhibits the outgrowth of axillary buds, a phenomenon termed 'apical dominance'. Removal of the shoot apex leads to the release of dormant axillary buds below it to form branches. Apical dominance allows plants to focus resources into the main axis of growth, while activation of dormant buds allows for recovery after damage or loss of the main shoot. It has been known for a long time that auxin is a major signal for apical dominance. If auxin is applied to a decapitated stump at the shoot apex, it can replace the growing apex in restoring branch inhibition [156]. Auxin is mainly synthesized in the shoot apex in young leaves [88] and subsequently transported basipetally (downwards from the tip to the base) in the polar auxin transport stream (PATS). It was shown that apical dominance depends on the PATS. For example, if auxin transport inhibitors such as 2,3,5-triiodobenzoic acid (TIBA) are applied to stems, the suppression of outgrowth is abolished [202]. Despite its long history, the mechanism by which auxin inhibits the growth of axillary meristems is not fully understood. However, it is clear that the auxin moving in the main stem must act indirectly because it does not move upward from the stem into the bud [203]. Therefore, a second messenger was proposed to regulate bud activity *in situ.*

25. Bud Activation and Auxin Transport

There is increasing evidence for a central role of auxin transport in the control of bud activity. There is a tight correlation between bud outgrowth and export of auxin from the bud [204]. For example, it has been shown that upon decapitation, auxin export from previously dormant pea buds is rapidly detected as an early event in bud activation. Such export is prevented by the application of auxin to the decapitated stump. The sustained export of auxin from buds requires auxin transport canalization. The canalization hypothesis was proposed by Sachs [205] to explain a wide range of phenomena involving the establishment of auxin transport pathways connecting auxin sources and auxin sinks. Core to canalization is a positive feedback loop between auxin flow and both the polarization and upregulation of active auxin transport. An initial auxin flow between a source and a sink upregulates auxin transport along the auxin transport pathway, culminating in the formation of narrow files of cells with high auxin transport activity from the source to the sink [104].

If sustained auxin export is required for bud activity, the canalization mechanism can explain the ability of auxin moving in the main stem to inhibit the activity of axillary buds indirectly. Auxin in the main stem would lower the sink strength of the stem and thus reduce the basal level of auxin flow between the bud and the main stem. If this level is too low to trigger the positive feedback of canalization, then the bud will remain dormant. Indeed application of apical auxin to the decapitated stump of peas is sufficient to prevent auxin transport canalization from the bud to the main stem as well as to inhibit the activation of the bud [204]. Thus, increased auxin export from the bud would reduce auxin levels in the bud and upregulate bud auxin biosynthesis. Computational modelling has shown that reduced feedback regulation such that auxin accumulates in the bud is alone sufficient to activate buds ectopically, because they become stronger auxin sources and are thus able to establish auxin transport into the stem even when it is a weak sink [206].

26. Abscission and Senescene

Several early studies revealed that exogenous application of auxin delayed leaf blade abscission in bean [207, 208]. Further, the concentration of free IAA in senescing leaves of Arabidopsis was 2-fold higher than in non-senescing leaves [209], which suggests either that auxin has a senescence-promoting effect or that it accumulates as a consequence of senescence.

Many studies have suggested the correlation between auxin levels with senescence and abscission [210]. One of them was reported that a gradient of auxin levels between the leaf blade and the stalk occurred in bean leaves. Auxin levels declined as leaf aged and senescence occurred when auxin levels between the leaf and stalk were approximately equal [211]. Recently, auxin was reported to repress transcription of some genes whose expression is correlated with senescence and/or abscission [212]. According to the literary data, auxin may retard [211] or accelerate [213] senescence in some cases in plants. Senescens is a genetically regulated metabolic breakdown process. During this process synthesis of hydrolytic enzymes increase and therefore molecules like chlorophyll and proteins breakdown quickly. Auxins may retard loss of proteins since they encourage synthesis of protease inhibitors.

Auxins have often been reported either to delay or to induce fruit abscission, and hence may operate as growth hormones or as abscising agents. On one hand, auxins promote cell enlargement rather than cell division. Although endogenous auxins also increase in developing ovaries, it is well established that exogenous treatments do not improve fruit set. Auxins are also high during the beginning of phase II, the period of cell elongation, and it is at this moment when exogenous auxins are effective increasing fruit size [214]. These observations may suggest that auxins are related to cell enlargement, the essential factor controlling fruit size during the phase of rapid growth. The enlargement of the auxin-treated fruits is apparently due to cell expansion rather than to cell division. In tomato, for example, it has been postulated that auxins are part of the hormonal signaling transduction network controlling cell expansion [215]. On the other hand, auxins may act either as delaying or accelerating agents of abscission. During the initial phases of abscission auxins operate as inhibitors, but once the process has been initiated auxins appear to stimulate abscission. Here, auxins could operate through the promotion of ethylene synthesis. It has also been suggested that the causal reason of the dual effect of auxins on leaf abscission may conveniently be explained by the auxin-gradient concept: auxin coming from the leaf would tend to delay abscission, whereas auxin moving down the stem might promote abscission. In citrus, synthetic auxins increase abscission of developing fruitlets during the cell division period. Auxin applications at the beginning of the cell enlargement period may have minor effects preventing fruit abscission and result in fruit size increase. The prevention or retardation effects of auxins on abscission can be perceived, however, at later stages. At the end of phase II or at the onset of phase III, synthetic auxins are commercially used to prevent or delay eventual pre-harvest fruit drop [216].

27. IAA

Indole-3-acetic acid (IAA) is the most common, naturally-occurring, plant hormone of the auxin class. It is the best known of the auxins, and has been the subject of extensive studies by plant physiologists. Chemically, IAA is a carboxylic acid in which the carboxyl group is attached through a methylene group to the C-3 position of an indole ring. In appearance, IAA is a colorless solid [217].

Properties of IAA	
Molecular formula	$C_{10}H_9NO_2$
Molar mass	175.184
Appearance	white solid
Melting point	168-170 °C (441-443 K)
Solubility in water	moderate

IAA is predominantly produced in cells of the apex (bud) and very young leaves of a plant. Plants can synthesize IAA by several independent biosynthetic pathways. Four of them start from tryptophan, but there is also a biosynthetic pathway independent of tryptophan [77]. Plants mainly produce IAA from tryptophan through indole-3-pyruvic acid [218, 219]. IAA is also produced from tryptophan through indole-3-acetaldoxime in Arabidopsis [220]. Chemically, IAA can be synthesized by the reaction of indole with glycolic acid in the presence of base at 250°C [221].

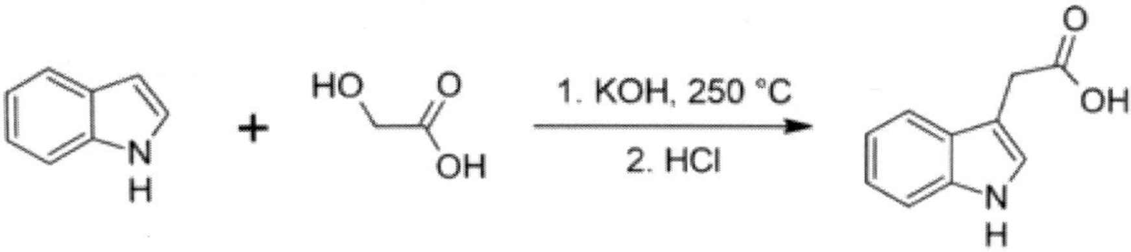

Many methods for its synthesis have been developed since its original synthesis from indole-3-acetonitrile [222].

28. IBA

Indole-3-butyric acid (1H-Indole-3-butanoic acid, IBA) is a white to light-yellow crystalline solid, with the molecular formula $C_{12}H_{13}NO_2$. It melts at 125 °C in atmospheric pressure and decomposes before boiling. IBA is a plant hormone in the auxin family and is an ingredient in many commercial horticultural plant rooting products.

Properties of IBA	
Molecular formula	$C_{12}H_{13}NO_2$
Molar mass	203.24 g mol^{-1}
Appearance	white to light yellow crystals
Density	1.252g/cm^3
Melting point	125 °C; 257 °F; 398 K
Boiling point	decomposes

This compound had been thought to be strictly synthetic; however, it was reported that the compound was isolated from leaves and seeds of maize and other species. In maize IBA has been shown to be synthesized in vivo using IAA and other compounds as precursors [223]. This chemical may also be extracted from any of the Salix (Willow) genus [224].

In plant tissue culture IBA and other auxins are used to initiate root formation *in vitro* in a procedure called micropropagation. Micropropagation of plants is the process of using small samples of plants called explants and causing them to undergo growth of differentiated or undifferentiated cells. In connection with cytokinins like kinetin, auxins like IBA can be used to cause the formation of masses of undifferentiated cells called *callus*. Callus formation is often used as a first step process in micropropagation where the callus cells are then caused to form other organs such as roots by exposing them to certain hormones like auxins that produce roots. The process of callus to root formation is called indirect organogenesis whereas if roots are formed from the explant directly it is called *direct organogenesis* [225].

Although the exact method of how IBA works is still largely unknown, genetic evidence has been found suggesting that IBA may be converted into IAA through a similar process to β-oxidation of fatty acids. The conversion of IBA to IAA suggesting that IBA works as a storage sink for IAA in plants [226]. There is other evidence that suggests that IBA is not converted to IAA but acts as an auxin on its own [223].

29. α-NAA

1-Naphthaleneacetic acid (NAA) is an organic compound with the formula $C_{10}H_7CH_2CO_2H$. This colorless solid is soluble in organic solvents. It features a carboxylmethyl group (CH_2CO_2H) linked to the "1-position" of naphthalene.

Properties of α-NAA (Dippy et al. 1954)	
Molecular formula	$C_{12}H_{10}O_2$
Molar mass	186.21 g mol^{-1}
Appearance	White powder
Melting point	135 °C (275 °F)
Solubility in water	0.38 g/L (17 °C)
Acidity (pK_a)	4.24

NAA is a synthetic plant hormone in the auxin family and is an ingredient in many commercial plant rooting horticultural products; it is a rooting agent and used for the vegetative propagation of plants from stem and leaf cuttings. It is also used for plant tissue culture [227].

NAA is widely used in agriculture for various purposes. It is considered to be only slightly toxic but when at higher concentrations it can be toxic to animals. This was shown when tested on rats via oral ingestion at 1000-5900 mg/kg [228]. NAA has been shown to greatly increase cellulose fiber formation in plants when paired with another phytohormone called gibberellic acid. Because it is in the auxin family it has also been understood to prevent premature dropping and thinning of fruits from stems. It is applied after blossom fertilization. Increased amounts of it can actually have negative effects however, and cause growth inhibition to the development of plant crops. It has been used on many different crops like apples, olives, oranges, potatoes, and various other hanging fruits. In order for it to obtain its desired effects it must be applied in concentrations ranging from 20-100 ug/mL [229]. NAA can be detected by HPLC-tandem mass spectrometry (HPLC-MS/MS) [230].

30. 2,4-D

Properties of 2,4-D (ChemIndustry site)	
Molecular formula	$C_8H_6Cl_2O_3$
Molar mass	221.04 g mol^{-1}
Appearance	white to yellow powder
Melting point	140.5 °C; 284.9 °F; 413.6 K
Boiling point	160 °C; 320 °F; 433 K (0.4 mm Hg)
Solubility in water	900 mg/L

2,4-Dichlorophenoxyacetic acid (usually referred to by its abbreviation, 2,4-D) is a common systemic pesticide/herbicide used in the control of broadleaf weeds. It is one of the most widely used herbicides in the world, and is the third most commonly used herbicide in North America (24d.org Industry Task Force II on 2,4-D Research Data). 2,4-D is a synthetic auxin (plant hormone), and as such it is often used in laboratories for plant research and as a supplement in plant cell culture media such as MS medium.

A number of 2,4-D-degrading bacteria have been isolated and characterized from a variety of environmental habitats [231, 232]. Metabolic pathways for the compound's degradation have been available for many years, and genes encoding 2,4-D catabolism have been identified for several organisms. As a result of the extensive metadata on environmental behavior, physiology and genetics, 2,4-D was the first herbicide for which the bacteria actively responsible for *in situ* degradation was demonstrated [233]. This was accomplished using the technique of DNA-based stable isotope probing, which enables a microbial function (activity), such as degrading a chemical, to be linked with the organism's identity without the need to culture the organism involved [234].

31. 2,4-D and Its Relation to Embryogenesis

Among different auxins, 2,4-dichlorophenoxyacetic acid (2,4-D) has been applied for induction of somatic embryogenesis the most commonly [235]. This synthetic growth regulator and auxinic herbicide appears to act not only as an exogenous auxin analogue but also as an effective stressor [235]. It has been hypothesized that 2,4-D concentration (around 10μM) acts at the same time as an auxin (either itself and/or through endogenous indole acetic acid (IAA) levels) and as a stressor [111]. Auxinic herbicides have been shown to interact with ethylene and abscisic acid (ABA) synthesis, increasing the cellular levels of these hormones [236, 237]. Rodrigues et al., [238] studied the effect of light and 2,4-D (2 and 10 mg/l) on the efficiency of microspore embryogenesis in soybean and found that 2,4-D at a higher concentration could not improve androgenesis. However, a high concentration (40 mg/l) of 2,4-D was employed to induce somatic embryogenesis from immature soybean cotyledons [239]. Also, Kim et al., [240] reported the induction of embryogenic calli (at the frequency of 53%) in petiole explants of *Hylomecon vernalis* when the explants were cultured on B5 medium supplemented with 13.6μM 2,4-D alone. Leaf protoplast-derived cells treated with 2,4-D could develop into pro-embryogenic cell clusters [241]. Sharma et al., [242] found out that the somatic embryogenesis in potato internodal segments could be evoked by 2,4-D pulse treatment over a wide concentration and duration range. 2,4-D also affects the activity/function of different genes responsible for biosynthesis pathways such as production of auxin, ABA and ethylene [243, 236, 237, 244, 245]. Perera et al., [246] investigated the effect of growth regulators on the efficiency of androgenesis in coconut. It was reported that 2,4-D (100μM) in combination with 1-naphthaleneacetic acid (NAA) (100μM) enhanced the production of calli/embryos. A high 2,4-D concentration (5–6 mg/l) increased embryo production in anther cultures of *Avena sterilis* and *A. sativa* × *A. sterilis* progeny and promoted plant regeneration in genotypes of both species [247].

2,4-D has been widely used for induction of embryogenesis in zygotic cells of *Arabidopsis thaliana* [248] and somatic cells in more species, e.g. *Hylomecon vernalis* [240], *Medicago sativa* [241], *Solanum tuberosum* [242]. 2,4-D has been employed in most anther/microspore culture systems either as an auxin (to elicit rapid cell proliferation and callus formation) or as a co-trigger for microspore differentiation in combination with other pre-treatments [249, 250, 251, 238, 252].

The incubation of wheat anthers with 2.0 or 4.0 mg/l 2,4-D in an induction medium for 10 or 15 days was reported to be sufficient for the initiation of callus development [251]. However, the reduction or removal of 2,4-D in the medium beyond the initiation phase was essential for plant regeneration from calli. It was also reported that the quality of auxin-free derived calli was poor. It is known that the first cell division during callus induction from anthers occurs within 2 days after inoculation for which 2,4-D plays a key role [253]. However, for the transition from callus and/or embryo development to plant regeneration, the reduction or removal of 2,4-D is needed since new gene products are required [251]. The same effect of 2,4-D as an auxin was reported in rice isolated microspore culture [249] in which 2,4-D was needed for induction of microspore-derived calli. Shariatpanahi et al., [252] also reported the positive influence of 2,4-D (2 mg/l in induction medium) on the quality of microspore-derived embryos in a wheat isolated microspore-culture system.

The influence of 2,4-D (10 mol/l continuously or for 1–7 days) in combination with mannitol pretreatment of barley anthers/isolated microspores was reported for induction of microspore differentiation [250]. They observed that without mannitol pretreatment no embryogenic type of microspores could be recognized at the moment of microspore isolation, and plating efficiency never reached 1%.

In another report, the combination of 2,4-D (10 mg/l) and light conditions (16h photoperiod at ±2000 lux) was shown to be an inducer of the callogenic response of anther walls and connective tissue in soybean anther culture, however the 2,4-D concentration had no effect on embryogenesis [238]. They reported plasmolysis of microspores in 10 mg/l 2,4-D both in the presence or absence of light, indicating a degradative effect of this auxin at this concentration. As explained above, there has been yet no report on the induction of microspore embryogenesis by 2,4-D alone. In this paper the use of 2,4-D alone as a trigger to induce microspore embryogenesis was investigated for the first time.

32. Concluding Remarks and Future Perspectives

Although our knowledge of auxin function and spatiotemporal distribution during reproductive development is still far from complete, some basic conclusions can be made from the data reviewed here. Auxin plays a fundamental role in the differentiation process in both male and female organs/tissues, and builds instructive gradients during polarity establishment (e.g., gynoecium and gametophyte). Auxin also appears to regulate the timing of events that requires coordination, such as anther dehiscence, pollen maturation, gynoecium development, fertilization, fruit initiation, seed development, fruit ripening, apical dominance, abscission, senescence, somatic embryogenesis, vascular differentiation, flower development, parthenocarpy, tropisms, and rooting. This is achieved through formation of local concentration maxima and minima, and most likely via interaction with other hormonal pathways. Identifying the mechanisms by which these interactions take place is one of the immediate challenges ahead for reproductive organ development.

Auxin biosynthesis in plants is complex and several pathways contribute to de novo IAA production. De novo auxin biosynthesis plays essential roles in virtually every aspect of plant development. Auxin biosynthesis is temporally and spatially regulated. Local auxin biosynthesis contributes to the generation and maintenance of local auxin gradients. Understanding of auxin biosynthesis provides new tools for solving difficult plant developmental questions. There is still no complete de novo auxin biosynthesis pathway in plants. The relationship among the proposed tryptophan dependent auxin biosynthesis pathways and the relative contribution of each pathway to overall IAA production need to be determined. Identification of the transcription factors that bind to the promoters of the key auxin biosynthesis genes will help to understand how auxin biosynthesis is regulated by developmental and environmental signals. Identification of the transcription factors that bind to the promoters of the key auxin biosynthesis genes will help to understand how auxin biosynthesis is regulated by developmental and environmental signals.

REFERENCES

[1] Darwin, C. R. (1880). *The Power of Movement in Plants*. London: Murray.

[2] Salkowski, E. (1885). "Uber das verhalten der skatolcarbonsaure im organismus". *Zeitschr. Physiol. Chem.*, *9*, 23-33.

[3] Arteca, R. (1996). Plant Growth Substances: Principles and Applications. New York, Chapman & Hall.

[4] Fitting, H. (1907). "Die Leitung Tropistischer Reize in parallelotropen Pflanzenteilen", *Jahrb. Wiss. Bot.*, *44*, 177-253.

[5] Boysen-Jensen, P. (1913). "Uber die Leitung des phototropischen Reizes in der Avenakoleoptile". *Ber. Deut. Bot. Ges.*, *31*, 559-566.

[6] Paal, A. (1918). "Uber phototropische Reizleitung". *Jahrb. Wiss. Bot.*, *58*, 406-458.

[7] Soding, H. (1925). "Zur kenntnis der wuchshormone in der haferkoleoptile". *Jahrb. Wiss. Bot.*, *64*, 587-603.

[8] Went, F. W. (1926). "On growth-accelerating substances in the coleoptile of *Avena sativa*". *Proc. Kon. Ned. Akad. Wet.*, *30*, 10-19.

[9] Went, F. W. (1928). "Wuchsstoff und Wachstum". *Rec. Trav. Bot. Neerland*, *24*, 1-116.

[10] Kogl, F. & Haagen-Smit, A. J. (1931). "Uber die Chemie des Wuchsstoffs K. Akad. Wetenschap. Amsterdam". *Proc. Sect. Sci.*, *34*, 1411-1416.

[11] Davies, P. J. (1995). Plant Hormones: Physiology, Biochemistry and Molecular Biology. Dordrecht: Kluwer.

[12] Laskowski, M., Biller, S., Stanley, K., Kajstura, T. & Prusty, R. (2006). Expression profiling of auxin-treated Arabidopsis roots: toward a molecular analysis of lateral root emergence. P*lant Cell Physiol*, *47*, 788–792. doi:10.1093/pcp/pcj043

[13] Ivanchenko, M. G., Coffeen, W. C., Lomax, T. L. & Dubrovsky, J. G. (2006). Mutations in the Diageotropica (Dgt) gene uncouple patterned cell division during lateral root initiation from proliferative cell division in the pericycle. *Plant J*, *46*, 436–447. doi:10.1111/j.1365-313X.2006. 02702.x

[14] Casimiro, I., Marchant, A., Bhalerao, R. P., Beeckman, T., Dhooge, S., Swarup, R., Graham, N., Inzé, D., Sandberg, G., Casero, P. J. & Bennett, P. (2001). Auxin transport promotes Arabidopsis lateral root initiation. *Plant Cell*, *13*, 843–852.

[15] Ljung, K., Hull, A. K., Celenza, J., Yamada, M., Estelle, M., Normanly, J. & Sandberg, G. (2005). Sites and regulation of auxin biosynthesis in Arabidopsis roots. *Plant Cell*, *17*, 1090–1104. doi:10.1105/tpc.104. 029272

[16] Swarup, R. & Bennett, M. (2003). Auxin transport: the fountain of life in plants? *Dev Cell*, *5*, 824–826 doi:10.1016/S1534-5807(03)00370-8

[17] Parry, G., Delbarre, A., Marchant, A., Swarup, R., Napier, R., Perrot-Rechenmann, C. & Bennett, M. J. (2001). Novel auxin transport inhibitors phenocopy the auxin influx carrier mutation aux1. *Plant J*, *25*, 399–406. doi:10.1046/j.1365-313x.2001.00970.x

[18] Lo Schiavo, F., Pitto, L., Ginliano, G., Torti, G., Nuti-Rouchi, V., Marazziti, D., Vergara, R., Orselli, S. & Terzi, M. (1989). DNA methylation of embryogenic carrot cell cultures and its variation as

caused by mutation, differentiation, hormones and hypomethylating drugs. *Theoretical and Applied Genetics*, *77*, 325-331.

[19] Ljung, K., Bhalerao, R. P. & Sandberg, G. (2002). Sites and homeostatic control of auxin biosynthesis in *Arabidopsis* during vegetative growth. *Plant J.*, *28*, 465-474.

[20] Baker, D. A. (2000). Long-distance vascular transport of endogenous hormones in plants and their role in source: sink regulation. *Israel J. Plant Sci.*, *48*, 199-203.

[21] Swarup, R., Friml, J., Marchant, A., Ljung, K., Sandberg, G., Palme, K. & Bennett, M. (2001). Localization of the auxin permease AUX1 suggests two functionally distinct hormone transport pathways operate in the *Arabidopsis* root apex. *Gene Dev.*, *15*, 2648-2653.

[22] Sundberg, B., Uggla, C. & Tuominen, H. (2000). Cambial growth and auxin gradients. In: Savidge, R.; Barnett, J.; Napier, R. (eds.) *Cell and Molecular Biology of Wood formation*. Oxford, 169-188.

[23] Srivastava, L. M. (2002). *Plant Growth and Development*, Academic Press, London.

[24] Muday, G. K. & DeLong, A. (2001). Polar auxin transport: controlling where and how much. *Trends Plant Sci.*, *6*, 535-542.

[25] Moyle, R., Schrader, J., Stenberg, A., Olsson, O., Saxena, S., Sandberg, G. & Bhalerao, R. P. (2002). Environmental and auxin regulation of wood formation involves members of the *Aux/IAA* gene family in hybrid aspen. *Plant J.*, *31*, 675-685.

[26] Uggla, C., Mellerowicz, E. J. & Sundberg, B. (1998). Indole-3-acetic acid controls cambial growth in Scots pine by positional signalling. *Plant Physiol.*, *117*, 113-121.

[27] Sundberg, B., Little, C. H. A. & Cui, K. (1990). Distribution of indole-3-acetic acid and the occurrence of its alkali-labile conjugates in the extraxylary region of *Pinus sylvestris* stems. *Plant Physiol.*, *93*, 1295-1302.

[28] Tuominen, H., Sitbon, F., Jacobsson, C., Sandberg, G., Olsson, O. & Sundberg, B. (1995). Altered growth and wood characteristics in transgenic hybrid aspen expressing *Agrobacterium tumefaciens* T-DNA indoleacetic acid-biosynthetic genes. *Plant Physiol.*, *109*, 1179-1189.

[29] Crozier, A., Loferski, K., Zaerr, J. B. & Morris, R. O. (1980). Analysis of picogram quantities of indole-3-acetic acid by high performance liquid chromatography-fluorescence procedures. *Planta*, *150*, 366-370.

[30] Little, C. H. A. & Savidge, R. A. (1987). The role of plant growth regulators in forest tree cambial growth. *Plant Growth Regul.*, *6*, 137-169.

[31] Little, C. H. A. & Pharis, R. P. (1995). Hormonal control of radial and longitudinal growth in the tree stem. In: Gartner, B. L. (ed). *Plant Stems: Physiology and Functional Morphology*. Academic Press, San Diego, CA, 281-319.

[32] Sundberg, B. & Little, C. H. A. (1990). Tracheid production in response to changes in the internal level of indole-3-acetic acid in 1-year-old shoots of Scots pine. *Plant Physiol.*, *94*, 1721-1727.

[33] Leitch, M. A. & Savidge, R. A. (1995). Evidence for auxin regulation of bordered-pit positioning during tracheid differentiation in *Larix laricina*. *IAWA*, *16*, 289-297.

[34] Kutschera, U. (1994). The current status of the acid-growth hypothesis. *New Phytol.*, *126*, 549-569.

[35] Roberts, L.W. (1988). Evidence from wound responses and tissue cultures. In: *Vascular Differentiation and Plant Growth Regulators*. Springer Verlag, Berlin, 63-88.

[36] Costacurta, A. & Vanderleyden, J. (1995). Synthesis of phytohormones by plant-associated bacteria. *Crit Rev Microbiol*, *21*, 1-18.

[37] Vandeputte, O., Oden, S., Mol, A., Vereecke, D., Goethals, K., El, J. M. & Prinsen, E. (2005). Biosynthesis of auxin by the gram-positive phytopathogen Rhodococcus fascians is controlled by compounds specific to infect plant tissues. *Appl. Environ. Microbiol.*, *71*, 1169-1177.

[38] Dobrev, P. I., Havlicek, L., Vagner, M., Malbeck, J. & Kaminek, M. (2005). Purification and determination of plant hormones auxin and abscisic acid using solid phase extraction and two-dimensional high performance liquid chromatography. *J. Chromatogr. A.*, *1075*, 159-166.

[39] Mohammad, G. & Prasad, R. (1998). Influence of microbial fertilizers on biomass accumulation in polypotted Eucalyptus calmadulensis seedlings. *J. Trop. For.*, *4*, 74-77.

[40] Muller, A., Duchting, P. & Weiler, E. W. (2002). A multiplex GC-MS/MS technique for the sensitive and quantitative single-run analysis of acidic phytohormones and related compounds, and its application to Arabidopsis thaliana. *Planta*, *216*, 44-56.

[41] Moore, T. C. (1979). Biochemistry and physiology of plants hormones. New York: Springer-Verlag.

[42] Normanly, J., Slovin, J. P. & Cohen, J. D. (1995). Rethinking auxin biosynthesis and metabolism. *Plant Physiol.*, *107*, 323-329.

[43] Guilfoyle, T., Hagen, G., Ulmasov, T. & Murfett, J. (1998). How does auxin turn on genes? *Plant Physiol.*, *118*, 341-347.

[44] Kim, Y. S., Min, J. K., Kim, D. & Jung, J. (2001). A soluble auxin-binding protein, ABP57. *J. Biol. Chem.*, *276*, 10730-10736.

[45] Friml, J. (2003). Auxin transport – shaping the plant. *Curr. Opin. Plant. Biol.*, *6*, 1-6. Auxins page on www.plant-hormones.info — a website set up at Long Ashton Research Station an institute of the BBSRC (site is now on independent server).

[46] Simon, S. & Petrášek, P. (2011). "Why plants need more than one type of auxin". *Plant Science*, *180*(3), 454–460. doi:10.1016/j.plantsci.2010. 12.007. PMID

[47] Taiz, L. & Zeiger, E. (1998). *Plant Physiology* (2nd ed.). Massachusetts: Sinauer Associates.

[48] Whippo, C. W. & Hangarter, R. P. (2006). "Phototropism: Bending towards enlightenment". *The Plant cell*, *18*(5), 1110–1119. doi:10.1105/ tpc.105.039669. PMC 1456868. PMID 16670442.

[49] Engvild, K. C. (1985). Pollen irradiation and possible gene transfer in *Nicotiana* species. *Theor. Appl. Genet.*, *69*, 457-461.

[50] Ludwig-Muller, J. & Epstein, E. (1991). Occurrence and *in vivo* biosynthesis of indole 3-butyric acid in corn (*Zea mays* L.). *Plant Physiol.*, 97, 765-770.

[51] Okamoto, I., Isogai, Y. & Koizumi, T. (1967). Isolation of indole-3-acetic acid, phenylacetic acid and several plant growth inhibitors from etiolated seedlings of *Phaseolus*. *Chem. Pharm. Bull.*, *15*, 159-163.

[52] Cooney, T. P. & Nonhebel, H. M. (1991). Biosynthesis of indole-3-acetic acid in tomato shoots – measurement, mass spectral identification and incorporation of H-2 from H20-H-2 into indole-3-acetic acid, D-tryptophan and L-tryptophan, indole-3-pyruvate and tryptamine. *Planta*, *184*, 368-376.

[53] Rayle, D. L. & Purves, W. K. (1967). Conversion of indole-3-ethanol to indole-3-acetic acid in cucumber seedling shoots. *Plant Physiol.*, *42*, 1091-1093.

[54] Percival, F. W., Purves, W. K. & Vickery, L. E. (1973). Indole-3-ethanol oxidase. Kinetics, inhibition and regulation by auxins. *Plant Physiol.*, *51*, 739-743.

[55] Saotome, M., Shirahata, K., Nishimura, R., Yahaba, M., Kawaguchi, M., Syono, K., Kitsuwa, T., Ishii, Y. & Nakamura, T. (1993). The identification of indole-3-acetic acid and indole-3-acetamide in the hypocotyls of Japanese cherry. *Plant Cell Physiol.*, *34*, 157-159.

[56] Sharma, A. K., Prasad, R. N. & Chaturvedi, H. C. (1981). Clonal propagation of *Bougainvillea glabra* 'Magnifica' through shoot apex culture. *Plant Cell Tiss. Organ Cult.*, *1*, 33-38.

[57] Guan, H. Y., Huisman, P. & De Klerk, G. J. (1997). Rooting of apple stem slices *in vitro* is affested by indoleacetic-acid depletion of the medium. *Angewandte Botanik*, *71*, 80-84.

[58] Davies, P. J. (2004). Regulatory factors in hormone action: Level, location and signal transduction. In: Davies, P. J. (ed.) *Plant Hormones*. Kluwer Academic Publishers, Dordrecht, 16-35.

[59] Zažímalová, E., Murphy, A. S., Yang, H., Hoyerová, K. & Hošek, P. (2010). Auxin transporters – why so many? *Cold Spring Harb. Perspect. Biol.*, *2*, a001552.

[60] Rosquete, M. R., Barbez, E. & Kleine-Vehn, J. (2012). Cellular auxin homeostasis: gatekeeping is housekeeping. *Mol. Plant*, *5*, 772-786.

[61] Law, D. M. & Davies, P. J. (1990). Comparative indole-3-acetic acid levels in the slender pea and other pea phenotypes. *Plant Physiol.*, *93*, 1539-1543.

[62] Pavlová, L. & Krekule, J. (1984). Fluctuations of free IAA under inductive and non-inductive photoperiods in *Chenopodium rubrum*. *Plant Growth Regul.*, *2*, 94-98.

[63] Cassells, A. C. (1979). The effect of 2,3,5-tri-iodobenzoic acid on caulogenesis in callus cultures of tomato and *Pelargonium*. *Physiol. Plant.*, *46*, 159-164.

[64] Fakhrai, H., Fakhrai, F. & Evans, P. K. (1989). *In vitro* culture and plant regeneration in *Vicia faba* subsp. *equina* (var. Spring Blaze). *J. Exp. Bot.*, *40*, 813-817.

[65] Clare, M. V. & Collin, H. A. (1974). The production of plantlets from tissue cultures of Brussels sprout (*Brassica oleracea* L. var. *gemmifera* D.C.). *Ann. Bot.*, *38*, 1067-1076.

[66] Forest, J. C. & Wightman, F. (1972). Amino acid metabolism in plants. III. Purification and some properties of a multispecific aminotransferase isolated from bushbean seedlings (*Phaseolus vulgaris* L.). *Can. J. Biochem.*, *50*, 813-829.

[67] Purves, W. K. & Brown, H. M. (1978). Indoleacetaldehyde in cucumber seedlings. *Plant Physiol.*, *61*, 104-106.

[68] Miyata, S., Suzuki, Y., Kamisaka, S. & Masuda, Y. (1981). Effects of abscisic acid on CO_2 exchange in *Lemna gibba*. *Physiol. Plant.*, *51*, 401-406.

[69] Normanly, J., Grisafi, P., Fink, G. R. & Bartel, B. (1997). Arabidopsis mutants resistant to the auxin effects on indole-3-acetonitrile are defective in the nitrilase encoded by the NIT1 gene. *Plant Cell*, *9*, 1781-1790.

[70] Kutáček, M. (1985). Auxin biosynthesis and its regulation on the molecular level. *Biol. Plant.*, *27*, 145-153.

[71] Klee, H. J., Horsch, R. B., Hinchee, M. A., Hein, M. B. & Hoffmann, N. L. (1987). The effects of overproduction of 2 *Agrobacterium tumefaciens* T-DNA auxin biosynthetic gene products in transgenic *Petunia* plants. *Genes & Develop.*, *1*, 86-96.

[72] Bartel, B. (1997). Auxin biosynthesis. *Annu. Rev. Plant Physiol. Plant Mol. Biol.*, *48*, 49-64.

[73] Patten, C. L. & Glick, B. R. (1996). Bacterial biosynthesis of indole-3-acetic acid. *Can. J. Microbiol.*, *42*, 207-220.

[74] Woodward, A. W. & Bartel, B. (2005). Auxin: regulation, action, and interaction. *Ann. Bot.*, *95*, 707–735.

[75] Chandler, J. W. (2009). Local auxin production: a small contribution to a big field. *Bioessays*, *31*, 60–70.

[76] Normanly, J. (2010). Approaching cellular and molecular resolution of auxin biosynthesis and metabolism. *Cold Spring Harbor Perspectives in Biology*, *2*, a001594.

[77] Zhao, Y. (2010). Auxin biosynthesis and its role in plant development. *Ann. Rev. Plant Biol.*, *61*, 49–64.

[78] Slovin, J. P., Bandurski, R. S. & Cohen, J. D. (1999). Auxin. In: Hooykaas, P. J. J., Hall, M. A., Libbenga, K. R. (eds). *Biochemistry and Molecular Biology of Plant Hormones.* Elsevier Science B.V., Amsterdam, 115-140.

[79] Klems, M., Truksa, M., Machackova, I., Eder, J. & Prochazka, S. (1998). Uptake, transport and metabolism of C^{14}-2,4-dichlorophenoxyacetic acid (C^{14}-2,4-D) in cucumber (*Cucumis sativus* L.) explants *Plant Growth Regul.*, *26*, 195-202.

[80] Zazimalova, E., Opatrny, Z., Brezinova, A. & Eder, J. (1995). The effect of auxin starvation on the growth of auxindependent tobacco cell culture: dynamics of auxin-binding activity and endogenous free IAA content. *J. Exp. Bot.*, *46*, 1205-1213.

[81] Ribnicky, D. M., Ilic, N., Cohen, J. D. & Cooke, T. J. (1996). The effects of exogenous auxins on endogenous indole-3-acetic acid metabolism-The implications for carrot somatic embryogenesis. *Plant Physiol.*, *112*, 549-558.

[82] Nissen, S. J. & Sutter, E. G. (1988). Stability of IAA and IBA in nutrient medium after autoclaving and after storage under various environmental conditions. *HortScience*, *23*, 758.

[83] Dunlap, J. R., Kresovich, S. & McGee, R. E. (1986). The effect of salt concentration on auxin stability in culture media. *Plant Physiol.*, *81*, 934-936.

[84] Hausman, J. F. (1993). Changes in peroxidase activity, auxin level and ethylene production during root formation by poplar shoots raised *in vitro*. *Plant Growth Regul.*, *13*, 263-268.

[85] Hopkins, W. G. (1995). *Introduction to Plant Physiology*. John Wiley & sons, New York, pp. 464.

[86] Raven, J. A. (1975). Transport of indoleacetic acid in plant cells in relation to pH and electrical potential gradients, and its significance for polar IAA transport. *New Phytol.*, *74*, 163-174.

[87] Morris, D. A., Friml, J. & Zazímalová, E. (2004). The transport of auxins. In: Davies, P. J. (ed). *Plant Hormones: Biosynthesis, Signal Transduction, Action!* Springer/Kluwer Acad. Publ., N.Y., Dordrecht, 437-470.

[88] Ljung, K., Ostin, A., Lioussanne, L. & Sandberg, G. (2001). Developmental regulation of indole-3-acetic acid turnover in Scots pine seedlings. *Plant Physiol.*, *125*, 464–475.

[89] Went, F. W. (1974). Reflections and speculations. *Annu. Rev. Plant Physiol.*, *25*, 1–26.

[90] Morris, D. A. & Thomas, A. G. (1978). A microautoradiographic study of auxin transport in the stem of intact pea seedlings (*Pisum sativum* L.). *J. Exp. Bot.*, *29*, 147–157.

[91] Nowacki, J. & Bandurski, R. S. (1980). Myo-inositol esters of indole-3-acetic acid as seed auxin precursors of *Zea mays L. Plant Physiol.*, *65*, 422–427.

[92] Cambridge, A. P. & Morris, D. A. (1996). Transfer of exogenous auxin from the phloem to the polar auxin transport pathway in pea (*Pisum sativum* L.). *Planta*, *199*, 583–588.

[93] Lomax, T. L., Muday, G. K. 7 Rubery, P. H. (1995). Auxin transport. In: Davies P. J. (ed.). *Plant Hormones: Physiology, Biochemistry and Molecular Biology.* Kluwer, Dordrecht, Netherlands, 509–530.

[94] Rashotte, A. M., Brady, S., Reed, R., Ante, S. & Muday, G. K. (2000). Basipetal auxin transport is required for gravitropism in roots of *Arabidopsis*. *Plant Physiol.*, *122*, 481–490.

[95] Goldsmith, M. H. M. (1977). The polar transport of auxin. *Annu. Rev. Plant. Physiol.*, *28*, 439– 478.

[96] Benning, C. (1986). Evidence supporting a model of voltage dependent uptake of auxin into cucurbita-pepo vesicles. *Planta*, *169*, 228–237.

[97] Katekar, G. F. & Geisler, A. E. (1977). Auxin transport inhibitors. *Plant Physiol.*, *60*, 826–829.

[98] Rubery, P. H. (1990). Phytotropins: receptors and endogenous ligands. *Symp. Soc. Exp. Biol.*, *44*, 119–146.

[99] Morris, D. A., Rubery, P. H., Jarman, J. & Sabater, M. (1991). Effects of inhibitors of protein synthesis on transmembrane auxin transport in *Cucurbita pepo* L. hypocotyl segments. *J. Exp. Bot.*, 42, 773–783.

[100] Cox, D. N. & Muday, G. K. (1994). NPA binding-activity is peripheral to the plasma membrane and is associated with the cytoskeleton. *Plant Cell*, *6*, 1941–1953.

[101] Bernasconi, P., Bhavesh, P. C., Reagan, J. D. & Subramanian, M. V. (1996). The N-naphthylphthalamic acid-binding protein is an integral membrane protein. *Plant Physiol.*, *111*, 427–432.

[102] Rayle, D. L. & Cleland, R. E. (1992). The acid growth theory of auxin-induced cell elongation is alive and well. *Plant Physiol.*, *99*, 1271-1274

[103] Aloni, R. (2004). The induction of vascular tissues by auxin. In: Davies, P. J. (ed.). *Plant Hormones*. Kluwer Academic Publishers, Dordrecht, 471-492.

[104] Sachs, T. (2000). Integrating cellular and organismic aspects of vascular differentiation. *Plant Cell Physiol.*, *41*, 649-656.

[105] Liu, C. M., Xu, Z. H. & Chua, N. H. (1993). Polar transport is essential for the establishment of bilateral symmetry during early plant embryogenesis. *Plant Cell*, *5*, 621-630.

[106] Hemerly, A., Ferreira, P., Engler, J., Engler, G., Inze, D. & Vanmontagu, M. (1993). The control of cell cycle in *Arabidopsis* plant cell cultures. *J. Plant Res.*, *3*, 51-56.

[107] Van Der Krieken, W. M., Breteler, H. & Visser, M. H. M. (1992). Uptake and metabolism of indolebutyric acid during root formation on *Malus* microcuttings. *Acta Bot. Neerl.*, *41*, 435-442.

[108] Cline, M. G. (1994). The role of hormones in apical dominance - new approaches to an old problem in plant development. *Physiol. Plant.*, *90*, 230-237.

[109] Terzi, M. & Lo Schiavo, F. (1990). Developmental mutants in carrot. In: Nijkamp, H. I. J.; Van Der Plas, I. H. W.; Van Aartrijk, J. (eds.). *Progress in Plant Cellular and Molecular Biology.* Proc. VIIth Int.

Cong. on Plant Tissue and Cell Culture. Amsterdam, The Netherlands. 24-29 June 1990. Kluwer Academic Publishers, Dortrecht, Netherlands, 391-397.

[110] Sharp, W. R., Sondahl, M. R., Caldas, L. S. & Maraffa, S. B. (1980). The physiology of *in vitro* asexual embryogenesis. *Hort. Rev.*, *2*, 268-310.

[111] Feher, A., Pasternak, T., Otvos, K., Miskolczi, P. & Dudits, D. (2002). Induction of embryogenic competence in somatic plant cells: A review. *Biologia*, *57*, 5-12.

[112] Bozhkov, P. V., Filonova, L. H. & Von Arnold, S. (2002). A key developmental switch during Norway spruce somatic embryogenesis is induced by withdrawal of growth regulators and is associated with cell death and extracellular acidification. *Biotechnol. Bioeng.*, *77*, 658-667.

[113] Hadfi, K., Speth, V. & Neuhaus, G. (1998). Auxin-induced developmental patterns in *Brassica juncea* embryos. *Development*, *125*, 879-887.

[114] Mattsson, J., Sung, Z. R. & Berleth, T. (1999). Responses of plant vascular systems to auxin transport inhibition. *Development*, *126*, 2691-2979.

[115] Schiavone, F. M. & Cooke, T. J. (1987). Unusual pattern of somatic embryogenesis in the domesticated carrot: Developmental effects of exogenous auxins and auxin transport inhibitors. *Cell Differ.*, *21*, 53-62.

[116] Morris, D. A. (2000). Transmembrane auxin carrier systems Ð dynamic regulators of polar auxin transport. *Plant Growth Reg.*, *32*, 161-172.

[117] Baker, C. M. & Wetzstein, H. Y. (1994). Influence of auxin type and concentration on peanut somatic embryogenesis. *Plant Cell Tiss. Org. Cult.*, *36*, 361-368.

[118] Ahmed, R., Gupta, S. D. & De, D. N. (1996). Somatic embryogenesis and plant regeneration from leaf derived callus of winged bean (*Psophocarpus tetragonolobus* L.). *Plant Cell Rep.*, *15*, 531-535.

[119] Eapen, S. & George, L. (1993). Somatic embryogenesis in peanut: Influence of growth regulators and sugars. *Plant Cell Tiss. Org. Cult.*, *35*, 151-156.

[120] Ozias-Akins, P., Anderson, W. F. & Holbrook, C. C. (1992). Somatic embryogenesis in *Arachis hypogaea* L.: genotype comparison. *Plant Sci.*, *83*, 103-111.

[121] Baker, C. M., Burns, A. & Wetzstein, H. Y. (1994). Influence of photoperiod and medium formulation on peanut somatic embryogenesis, *Plant Cell Rep.*, *13*, 159-163.

[122] Hazra, S., Sathaye, S. S. & Mascarenhas, A. F. (1989). Direct somatic embryogenesis in peanut (*Arachis hypogaea* L). *BioTechnol.*, *7*, 949-951.

[123] Ozias-Akins, P. (1989). Plant regeneration from immature embryos of peanut. *Cell Rep.*, *8*, 217-218.

[124] Sellars, R. M., Southward, G. M. & Phiillips, G. C. (1990). Adventitious somatic embryogenesis from cultured immature zygotic embryos of peanut and soybean. *Crop Sci.*, *30*, 408-414.

[125] Mckently, A. H. (1991). Direct somatic embryogenesis from axes of mature peanut embryos. *In Vitro Cell Dev. Biol.*, *27P*, 197-200.

[126] Venkatachalam, P., Kavi Kishor, P. B. & Jayabalan, N. (1997). High frequency somatic embryogenesis and efficient plant regeneration from hypocotyls explants of groundnut (*Arachis hypogaea* L.). *Curr. Sci.*, *72*, 271-275.

[127] Liu, C., Xu, Z. & Chua, N. H. (1993). Auxin polar transport is essential for the establishment of bilateral symmetry during early plant embryogenesis. *Plant Cell*, *5*, 621-630.

[128] Minocha, S. C. (1987). pH of the medium and the growth and metabolism of cells in culture. In: Bonga and Durzan (eds.). *Cell and Tissue Culture in Forestry.* Vol *1. General Principles and Biotechnology*. Martinus Nijhoff Publishers, Dordrecht, Boston, Lancaster, 125-141. ISBN 90-247-3430-4.

[129] Sachs, T. (1991). *Pattern Formation in Plant Tissues*. Cambridge University Press, Cambridge.

[130] Hobbie, L. J. (1998). Auxin: molecular genetic approaches in *Arabidopsis*. *Plant Physiol. Biochem.*, *36*, 91-102.

[131] Boiero, L., Perrig, D., Masciarelli, O., Penna, C., Cassán, F. & Luna, V. (2007). Phytohormone production by three strains of *Bradyrhizobium japonicum* and possible physiological and technological implications. *Appl. Microbiol. Biotechnol.*, *74*, 874-880.

[132] Bianco, C. & Defez, R. (2009). *Medicago truncatula* improves salt tolerance when nodulated by an indole-3-acetic acid overproducing *Sinorhizobium meliloti* strain. *J. Exp. Bot.*, *60*(11), 3097-3107.

[133] Wang, Y., Mopper, S. & Hasentein, K. H. (2001). Effects of salinity on endogenous ABA, IAA, JA, and SA in *Iris hexagona*. *J. Chem. Ecol.*, *27*, 327-342.

[134] Kende, H. & Zeevaart, J. A. D. (1997). The five "classical" plant hormones. *Plant Cell*, *9*, 1197–1210.

[135] Sterling, T. M. & Hall, J. C. (1997). Mechanism of action of natural auxins and the auxinic herbicides. In: Roe, R. M.; Burton, JD.; Kuhr, RJ. (eds.). *Herbicide Activity: Toxicology, Biochemistry and Molecular Biology*. IOS Press, Amsterdam, 111–141.

[136] Grossmann, K. (1998). Quinclorac belongs to a new class of highly selective auxin herbicides. *Weed Sci.*, *46*, 707–716.

[137] Klee, H. J. & Lanahan, M. B. (1995). Transgenic plants in hormone biology. In: Davies, PJ. (ed.). *Plant Hormones: Physiology, Biochemistry and Molecular Biology*. Kluwer Academic Publishers, Dordrecht, The Netherlands, 340–353.

[138] Lincoln, C., Britton, J. H. & Estelle, M. (1990). Growth and development of the axr1 mutants of *Arabidopsis*. *Plant Cell*, *2*, 1071–1080.

[139] Chen, R., Guan, C., Boonsirichai, K. & Masson, P. H. (2002). Complex physiological and molecular processes underlying root gravitropism. *Plant Mol. Biol.*, *49*, 305–317.

[140] Ward, S. P. & Leyser, O. (2004). Shoot branching. *Curr. Opin. Plant Biol.*, *7*, 73–77.

[141] Masucci, J. D. & Schiefelbein, J. W. (1994). The rhd6 mutation of Arabidopsis thaliana alters root-hair initiation through an auxin-associated and ethylene-associated process. *Plant Physiol.*, *106*, 1335–1346.

[142] Skoog, F. & Miller, C. O. (1957). Chemical regulation of growth and organ formation in plant tissues cultured *in vitro*. *Symp. Soc. Exp. Biol.*, *11*, 118–130.

[143] Malamy, J. (2005). Intrinsic and environmental response pathways that regulate root system architecture. *Plant Cell Environ.*, *28*, 67-77.

[144] Dubrovsky, J. G., Sauer, M., Napsucialy-Mendivil, S., Ivanchenko, M. G., Friml, J., Shishkova, S., Celenza, J. & Benkova, E. (2008). Auxin acts as a local morphogenetic trigger to specify lateral root founder cells. *Proc. Natl. Acad. Sci. U.S.A.*, *105*, 8790-8794.

[145] De-Klerk, G., Krieken, W. & DeJong, J. (1999). The formation of adventitious roots: New concepts, new possibilities. *In Vitro Cell Dev Biol-Plant*, *35*, 189-199.

[146] Riov, J. & Yang, S. (1989). Ethylene and Auxin-ethylene interaction in adventitious root formation in Mung bean (*Vigna radiata*) cuttings. *J. Plant Growth Regul.*, *8*, 131-141.

[147] Benková, E., Michniewicz, M., Sauer, M., Teichmann, T., Seifertová, D., Jürgens, G. & Friml, J. (2003). Local, efflux-dependent auxin

gradients as a common module for plant organ formation. *Cell, 115*(5), 591–602. doi:10.1016/S0092-8674(03)00924-3. PMID 14651850

[148] Mauseth, J. D. (1991). Botany: An Introduction to Plant Biology. Philadelphia: Saunders. 348-415.

[149] Raven, P. H., Evert, R. F. & Eichhorn, S. E. (1992). *Biology of Plants*. New York: Worth. 545-572.

[150] Salisbury, F. B. & Ross, C. W. (1992). Plant Physiology. Belmont, CA: Wadsworth. 357-407, 531-548.

[151] Martin, G. C. (1987). Apical dominance. *HortScience*, *22*, 824-833.

[152] Cline, M. G. (1991). Apical dominance. *The Botanical Review*, *57*, 318-358.

[153] Bangerth, F. (1994). Response of cytokinin concentration in the xylem exudate of bean plants (*Phaseolus vulgaris* L.) to decapitation and auxin treatment and relationship of apical dominance. *Planta*, *194*, 439-442.

[154] Stafstrom, J. (1995). Developmental potential of shoot buds. In: Gartner, B. L. (ed.). *Plant stems: Physiology and functional morphology*. San Diego: Academic Press, 257-279.

[155] Tamas, I. A. (1995). Hormonal regulation of apical dominance. In: Davies, P. J. (ed.). *Plant hormones, physiology, biochemistry and molecular biology*. 2nd edn. Boston: Kluwer Academic Publishers, 572-597.

[156] Thimann, K. V. & Skoog, F. (1933). Studies on the growth hormones of plants. III. The inhibition action of growth substance on bud development. *Proceedings of the National Academy of Science*, *USA*, *19*, 714-716.

[157] Mullen, J. L. & Kiss, J. Z. (2008). Phototropism and its relationship to gravitropism. In: Gilroy, S.; Masson, P. H. (eds.). *Plant Tropisms*. Blackwell Publishing Ltd., Ames, Iowa, 79-90.

[158] Muday, G. K. & Rahman, A. (2008). Auxin transport and the integration of gravitropic growth. In: Gilroy, S.; Masson, P. H. (eds.). *Plant Tropisms*. Blackwell Publishing Ltd., Ames, Iowa, 47–77.

[159] Yamamoto, K. T. (2003). Happy end in sight after 70 years of controversy. *Trends Plant Sci.*, *8*, 359–360.

[160] Hager, A. (2003). Role of the plasma membrane H+-ATPase in auxin-induced elongation growth: historical and new aspects. *J. Plant Res.*, *116*, 483–505.

[161] Grebe, M. (2005). Growth by auxin: when a weed needs acid. *Science*, *310*, 60–61.

[162] Philosoph-Hadas, S., Friedman, H., Meir, S. & Gerald, L. (2005). Gravitropic bending and plant hormones. *Vitam. Horm.*, *72*, 31–78.

[163] Buer, C. S., Sukumar, P. & Muday, G. K. (2006). Ethylene modulates flavonoid accumulation and gravitropic responses in roots of Arabidopsis. *Plant Physiol.*, *140*, 1384–1396.

[164] Shen-Miller, J., Cooper, P. & Gordon, S. A. (1969). Phototropism *and photoinhibition of basipolar transport of auxin in oat coleoptiles. Plant Physiol., 44, 491–496.*

[165] Naqvi, S. M. (1972). *Possible role of abcissic acid in phototropism. Z. Pflanzenphysiol., 67, 454–456.*

[166] Gardner, G., Shaw, S. & Wilkins, M. B. (1974). IAA *transport during the phototropic responses of intact Zea and Avena coleoptiles. Planta, 28,* 439–*478.*

[167] Kang, B. G. & Burg, S. P. (1974). *Red Light enhancement of the phototropic response in pea stems. Plant Physiol.*, *53*, 445–448.

[168] Cheng, Y., Dai, X. & Zhao, Y. (2006). Auxin biosynthesis by the YUCCA flavin monooxygenases controls the formation of floral organs and vascular tissues in *Arabidopsis*. *Genes Dev.*, *20*, 1790–1799.

[169] Heisler, M., Ohno, C., Das, P., Sieber, P., Reddy, G. V., Long, J. A. & Meyerowitz, E. M. (2005). Patterns of auxin transport and gene expression during primordium development revealed by live imaging of the Arabidopsis inflorescence meristem. *Curr. Biol.*, *15*, 1899–1911.

[170] Kaufmann, K., Muin, J. M., Jauregui, R., Airoldi, C. A., Smaczniak, C., Krajewski, P. & Angenent, G. C. (2009). Target genes of the MADS transcription factor SEPALLATA3: Integration of developmental and hormonal pathways in the Arabidopsis flower. *PLOS Biol.*, *7*, e1000090.

[171] Benkova,´ E., Michniewichz, M., Sauer, M., Tiechmann, T., Seifertova,´ D., Jurgens, G. & Friml, J. (2003). Local, efflux dependent auxin gradients as a common module for plant organ formation. *Cell*, *115*, 591–602.

[172] Cheng, Y. & Zhao, Y. (2007). A role for auxin in flower development. *J. Integr. Plant Biol.*, *49*, 99–104.

[173] Sundberg, E. & Ostergaard, L. (2009). Distinct and dynamic auxin activities during reproductive development. *Cold Spring Harbor Perspectives in Biology*, *1*, a001628.

[174] Cecchetti, V., Altamura, M. M., Falasca, G., Constantino, P. & Cardarelli, M. (2008). Auxin regulates Arabidopsis anther dehiscence, pollen maturation, and filament elongation. *Plant Cell*, *20*, 1760–1774.

[175] Pacini, E., Franchi, G. G. & Hesse, M. (1985). The tapetum: Its form, function and possible phylogeny in embryophyta. *Plant System Evol.*, *149*, 155–185.

[176] Aloni, R., Aloni, E., Langhans, M. & Ullrich, C. I. (2006). Role of auxin in regulating Arabidopsis flower development. *Planta*, *223*, 315–328.

[177] Yang, H., Wei, Z. & Xu, Z. (1997). Effects of specific expression of iaaL in tobacco tapetum on pollen embryogenesis. *Sci. China C. Life Sci.*, *40*, 384–391.

[178] Feng, X. L., Ni, W. M., Egle, S., Mueller-Roeber, B., Xu, Z. H. & Xue, H. W. (2006). Auxin flow in anther filaments is critical for pollen grain development through regulating pollen mitosis. *Plant Mol. Biol.*, *61*, 215–226.

[179] Cheng, Y., Dai, X. & Zhao, Y. (2006). Auxin biosynthesis by the YUCCA flavin monooxygenases controls the formation of floral organs and vascular tissues in *Arabidopsis. Genes Dev.*, *20*, 1790–1799.

[180] Cantliffe, D. J. (1972). Parthenocarpy in the cucumber induced by some plant growth-regulating chemicals. *Can. J. Plant Sci.*, *52*, 781-785.

[181] Cantliffe, D. J., Robinson, R. W. & Bastdorff, R. S. (1972). Parthenocarpy of cucumber induced by triiodobenzoic acid. *HortScience*, *7*, 285-286.

[182] Quebedeaux, B. & Beyer, E. M. (1972). Chemically induced parthenocarpy in the cucumber by a new inhibitor of auxin transport. *HortScience*, *7*, 474-476.

[183] Robinson, R., Cantliffe, D. J. & Shannon, S. (1971). Morphactin induced parthenocarpy in the cucumber. *Science*, *171*, 1251-1252.

[184] Gttstafson, F. G. (1939). The cause of natural parthenocarpy. *Am. J. Bot.*, *26*, 135-138.

[185] Beyer, E. M. & Quebedeaux, J. R. (1974). Parthenocarpy in cucumber: Mechanism of action of auxin transport inhibitiors. *J. Am. Soc. Hort. Sci.*, *99*, 385-390.

[186] Elassar, G., Rudich, J., Palevitch, D. & Kedar, N. (1974). Induction of parthenocarpic fruit development in cucumbers by growth regulators. *HortScience*, *9*, 23-239.

[187] Catala,´C., Rose, J. K. & Bennett, A. B. (2000). Auxin-regulated genes encoding cell wall-modifying proteins are expressed during early tomato fruit growth. *Plant Physiol.*, *122*, 527–534.

[188] Aharoni, A., Keizer, L. C., Van Den Broeck, H. C., Blanco-Portales, R., Munoz-Blanco, J., Bois, G., Smit, P., DeVos, R. C. & O'Connell, A. P. (2002). Novel insight into vascular, stress, and auxin-dependent and -

independent gene expression programs in strawberry, a non-climacteric fruit. *Plant Physiol., 129*, 1019–1031.

[189] Aloni, R. (1987). Differentiation of vascular tissues. *Annu. Rev. Plant Physiol., 38*, 179–204.

[190] D'Agostino, I. B. & Kieber, J. J. (1999). Molecular mechanisms of cytokinin action. *Curr. Opin. Plant Biol., 2*, 359–64.

[191] den Boer, B. G. W. & Murray, J. A. H. (2000). Triggering the cell cycle in plants. *Trends Cell Biol., 10*, 245–50.

[192] Uggla, C., Moritz, T., Sandberg, G. & Sundberg, B. (1996). Auxin as a positional signal in pattern formation in plants. *Proc. Natl. Acad. Sci. USA, 93*, 9282–86.

[193] Aloni, R. (1993). The role of cytokinin in organised differentiation of vascular tissues. *Aust. J. Plant Physiol., 20*, 601–608.

[194] Romano, C. P., Hein, M. B. & Klee, H. J. (1990). Inactivation of auxin in tobacco transformed with the indoleacetic acid-lysine synthetase gene of *Pseudomonas savastanoi*. *Gene Dev., 5*, 438–46.

[195] Li, C., Junttila, O., Ernstsen, A., Heino, P. & Palva, E. T. (2003). Photoperiodic control of growth, cold acclimation and dormancy development in silver birch (*Betula pendula*) ecotypes. *Physiol. Plant., 117*, 206-212.

[196] Sorce, C., Lorenzi, R., Ceccarelli, N. & Ranalli, P. (2000). Changes in free and conjugated IAA during dormancy and sprouting of potato tubers. *Aust. J. Plant Physiol., 27*, 371-377.

[197] Chao, W. S., Serpe, M. D., Anderson, J. V., Gesch, R. W. & Horvath, D. P. (2006). Sugars, hormones, and environment affect the dormancy status in underground adventitious buds of leafy spurge (*Euphorbia esula*). *Weed Sci., 54*, 59-68.

[198] Wan, X., Landhausser, S. M., Lieffers, V. J. & Zwiazek, J. J. (2006). Signals controlling root suckering and adventitious shoot formation in aspen (*Populus tremuloides*). *Tree Physiol., 26*, 681-687.

[199] Cline, M. G. (1997). Concepts and terminology of apical dominance. *Am. J. Bot., 84*, 1064-1069.

[200] Dun, E. A., Ferguson, B. J. & Beveridge, C. A. (2006). Apical dominance and shoot branching. Divergent opinions or divergent mechanisms? *Plant Physiol., 142*, 812-819.

[201] Bennett, T., Sieberer, T., Willett, B., Booker, J., Luschnig, C. & Leyser, O. (2006). The Arabidopsis MAX pathway controls shoot branching by regulating auxin transport. *Curr. Biol., 16*, 553-563.

[202] Panigrahi, B. M. & Audus, L. J. (1966). Apical dominance in *Vicia faba*. *Ann. Bot.*, *30*, 457–473.

[203] Booker, J., Chatfield, S. & Leyser, O. (2003). Auxin acts in xylem-associated or medullary cells to mediate apical dominance. *Plant The Cell*, *15*, 495–507.

[204] Balla, J., Kalousek, P., Reinohl, V., Friml, J. & Prochazka, S. (2011). Competitive canalization of PIN-dependent auxin flow from axillary buds controls pea bud outgrowth. *The Plant J.*, *65*, 571–577.

[205] Sachs, T. (1981). The control of the patterned differentiation of vascular tissues *Adv. Bot. Res.*, *9*, 151–262.

[206] Prusinkiewicz, P., Crawford, S., Smith, R. S., Ljung, K., Bennett, T., Veronica Ongaro, V. & Leyser, O. (2009). Control of bud activation by an auxin transport switch. *Proc. Nat. Acad. Sci. USA*, *106*, 17431–17436.

[207] Shoji, K., Addicott, F. T. & Swets, W. A. (1951). Auxin in relation to leaf blade abscission. *Plant Physiol*., *26*, 189–191.

[208] Sacher, J. A. (1957). Relationship between auxin and membrane-integrity in tissue senescence and abscission. *Science*, *125*, 1199–1200.

[209] Quirino, B., Normanly, J. & Amasino, R. (1999). Diverse range of gene activity during Arabidopsis thaliana leaf senescence including pathogen-independent induction of defense-related genes. *Plant Mol. Biol.*, *40*, 267–278.

[210] Nooden, L. D. & Leopold, A. (1998). The phenomena of senescence and aging. In: *Senescence and Aging in Plants*, Academic Press, San Diego, 1-50.

[211] Ellis, C. M., Nagpal, P., Young, J. C., Hagen, G., Guilfoyle, T. J. & Reed, J. W. (2005). Auxin response factor 1 and auxin response factor 2 regulate senescence and floral organ abscission in *Arabidopsis thaliana*. *Development*, *132*, 4563-4574.

[212] Tucker, M. L., Whitelav, C. A., Lyssenko, N. N. & Nath, L. (2002). Functional analysis of regulatory elements in the gene promoter for abscission-specific cellulose from bean and isolation, expression, and binding affinity of three TGA-type basic leucine zipper transcription factors. *Plant Physiol.*, *130*, 1487-1496.

[213] Aharoni, N. (1989). Interrelationship between ethylene and growth regulators in the senescence of lettuce leaf discs. *J. Plant Grow. Regul.*, *8*, 309-317.

[214] Coggins, C. W. & Hield, H. Z. (1968). Plant-growth regulators. In: Reuther, W.; Batchelor, L. D., Webber, H. J. (eds.). *The Citrus Industy.* Vol. 11. University of California Press, Berkeley, 371-389.

[215] Catala, C., Rose, J. K. C. & Bennett, A. B. (2000). Auxin-regulated genes encoding cell wall-modifying proteins are expressed during early tomato fruit growth. *Plant Physiol.*, *122*, 527-534.

[216] Agustí, M., Zaragoza, S., Iglesias, D.J., Almela, V., Primo-Millo, E. & Talon, M. (2002). The synthetic auxin 3,5,6-TPA stimulates carbohydrate accumulation and growth in citrus fruit. *Plant Growth Regul.*, *36*, 43-49.

[217] Simon, S. & Petrášek, J. (2011). Why plants need more than one type of auxin. *Plant Science* 180(3), 454–460. doi:10.1016/j.plantsci.2010.12.007. PMID 21421392.

[218] Mashiguchi, K., Tanaka, K., Sakai, T., Sugawara, S., Kawaide, H., Natsume, M., Hanada, A., Yaeno, T., et al. (2011). The main auxin biosynthesis pathway in Arabidopsis. *Proc. Nat. Acad. Sci.*, *108*(45), 18512–185177. doi, 10.1073/pnas.1108434108.

[219] Won, C., Shen, X., Mashiguchi, K., Zheng, Z., Dai, X., Cheng, Y., Kasahara, H., Kamiya, Y., et al. (2011). Conversion of tryptophan to indole-3-acetic acid by tryptophan aminotransferases of *Arabidopsis* and *Yuccas* in *Arabidopsis*. *Proc. Nat.l Acad. Sci.*, *108*(45), 18518–18523. doi:10.1073/pnas.1108436108.

[220] Sugawara, S., Hishiyama, S., Jikumaru, Y., Hanada, A., Nishimura, T., Koshiba, T., Zhao, Y., Kamiya, Y., et al. (2009). Biochemical analyses of indole-3-acetaldoxime-dependent auxin biosynthesis in Arabidopsis. *Proc. Nat. Acad. Sci.*, *106*(13), 5430–5435. doi:10.1073/pnas.0811226106.

[221] Johnson, H. E. & Crosby, D. G. (1964). Indole-3-acetic Acid. *Org. Synth.*, *44*, 64, *Coll.*, Vol., *5*, 654

[222] Majima, R. & Hoshino, T. (1925). Synthetische Versuche in der Indol-Gruppe, VI.: Eine neue Synthese von β-Indolyl-alkylaminen. *Berichte der deutschen chemischen Gesellschaft (A and B Series)*, *58*(9), 2042–2046. doi:10.1002/cber.19250580917

[223] Ludwig-Müller, J. (2000). Indole-3-butyric acid in plant growth and development. *Plant Growth Regul.*, *32*(2-3).

[224] William, G. H. & Norman, P. A. H. (1999). *Introduction to plant physiology.* Wiley. ISBN 978-0-471-19281-7.

[225] Bridgen, M. P., Masood, Z. H. & Spencer-Barreto, M. (1992). A laboratory exercise to demonstrate direct and indirect shoot

organogenesis from leaves of *Torenia fournieri*. *HortTechnology*., *4*(3), 320–322.

[226] Zolman, B. K., Martinez, N., Millius, A., Adham, A. R. & Bartel, B. (2008). Identification and characterization of Arabidopsis indole-3-butyric acid response mutants defective in novel peroxisomal enzymes. *Genitics*, *180*(1).

[227] Morikawa, H. & Takahashi, M. (2002). Cultured cells of Australian laurel, Pittosporaceae and a method for culturing tissues by using said cultured cells. issued 2004-10-05

[228] Tomlin, C. D. S. (2006). The Pesticide Manual. 14th ed. UK.

[229] Navalσn, A., Blanc, R. & Vilchez, J. L. (1997). Determination of 1-naphthylacetic acid in commercial formulations and natural waters by solid-phase spectrofluorimetry Mikcrochim.

[230] Benotti, M. J., Lee, F. P., Rieger, R. A., Iden, C. R., Heine, C. E. & Brownawell, B. J. (2003). HPLC/TOF-MS: an alternative to LC/MS/MS for sensitive and selective determination of polar organic contaminants in the aquatic environment. In: Ferrer, I.; Thurman, M. E. (eds.). Liquid Chromatography/Mass Spectrometry, MS/MS and Time of Flight MS. Am. Chem. Soc., New York, 109–127.

[231] Cavalca, L., Hartmann, A., Rouard, N. & Soulas, G. (1999). Diversity of tfdC genes: distribution and polymorphism among 2,4-dichlorophenoxyacetic acid degrading soil bacteria. *FEMS Microbiol. Ecol.*, *29*, 45-58.

[232] Suwa, Y., Wright, A. D., Fukimori, F., Nummy, K. A., Hausinger, R. P., Holben, W. E. & Forney, L. J. (1996). Characterization of a chromosomally encoded 2,4-dichlorophenoxyacetic acid alpha-ketoglutarate dioxygenase from Burkholderia sp. strain RASC". *Appl. Environ. Microbiol.*, *62*, 2464–2469.

[233] Cupples, A. M. & Sims, G. K. (2007). Identification of *in situ* 2,4-dichlorophenoxyacetic acid-degrading soil microorganisms using DNA-stable isotope probing. *Soil Biol. Biochem.*, *39*, 232–238.

[234] Radajewski, S., Ineson, P., Parekh, N. R. & Murrell, J. C. (2000). Stable-isotope probing as a tool in microbial ecology. *Nature*, *403*, 646–649.

[235] Gaj, M. D. (2004). Factors influencing somatic embryogenesis induction and plant regeneration with particular reference to *Arabidopsis thaliana* L. Heynh. *Plant Growth Regul.*, *43*, 27–47.

[236] Grossmann, K. (2000). Mode of action of auxin herbicides: a new ending to a long, drawn out story. *Trends Plant Sci.*, *5*, 506–508.

[237] Wei, Y. D., Zheng, H. G. & Hall, J. C. (2000). Role of auxinic herbicide-induced ethylene on hypocotyl elongation and root/hypocotyl radial expansion. *Pest Management Sci.*, *56*, 377–387.

[238] Rodrigues, L. R., Forte, B. D. C., Oliveira, J. M. S., Mariath, J. E. A. & Bodanese-Zanettini, M. H. (2004). Effects of light conditions and 2,4-D concentration in soybean anther culture. *Plant Growth Regul.*, *44*, 125–131.

[239] Bailey, M. A., Boerma, H. R. & Parrot, W. A. (1993). Genotypes effects on proliferative embryogenesis and plant regeneration of soybean. *In Vitro Cell. Dev. Biol.*, *29*, 102–108.

[240] Kim, S. W., In, D. S., Kim, T. J. & Liu, J. R. (2004). High frequency somatic embryogenesis and plant regeneration in petiole and leaf explants cultures and petiole-derived embryogenic cell suspention cultures of *Hylomecon vernalis*. *Plant Cell, Tiss. Org. Cult.*, *74*, 163–167.

[241] Pasternak, T., Prinsen, E., Ayaydin, F., Miskolezi, P., Potters, G., Asard, H., Van Onckelen, H., Dudits, D. & Feher, A. (2002). The role of auxin, pH and stress in the activation of embryogenic cell division in leaf protoplast-derived cells of alfalfa (*Medicago sativa* L). *Plant Physiol.*, *129*, 1807–1819.

[242] Sharma, S. K., Bryan, G. J. & Millam, S. (2007). Auxin pulse treatment holds the potential to enhance efficiency and practicability of somatic embryogenesis in potato. *Plant Cell Rep.*, *26*, 945–950.

[243] Michalczuk, L., Ribnicky, D. M., Cook, T. J. & Cohen, J. D. (1992). Regulation of indole-3-acetic acid biosynthetic pathways in carrot cell cultures. *Plant Physiol.*, *100*, 1346–1353.

[244] Feher, A., Pasternak, T. P. & Dudits, D. (2003). Transition of somatic plant cells to an embryogenic state. *Plant Cell, Tiss. Org. Cult.*, *74b*, 201–228.

[245] Raghavan., Ong E. K., Dalling, M. J. & Stevenson, T. W. (2006). Regulation of genes associated with auxin, ethylene and ABA pathways by 2,4-dichlorophenoxyacetic acid in *Arabidopsis*. *Functional and Integrative Genomics*, *6*, 60–70.

[246] Perera, P., Yakandawala, D., Hocher, V., Verdeil, J. L. & Weerakoon, L. (2009). Effect of growth regulators on microspore embryogenesis in coconut anthers. *Plant Cell Tiss. Org. Cult.*, *96*, 171–180.

[247] Kiviharju, E. M. & Tauriainen, A. A. (1999). 2,4-Dichlorophenoxyacetic acid and kinetin in anther culture of cultivated and wild

oats and their interspecific crosses: plant regeneration from *A. sativa* L. *Plant Cell Rep.*, *18*, 7–8.

[248] Raghavan, V. (2005). Control of leaf formation and somatic embryogenesis in cultured zygotic embryos of *Arabidopsis thaliana* by 2,4-dichlorophenoxyacetic acid (2,4-D). *Int. J. Plant Sci.*, *166*, 575–589.

[249] Xie, J., Gao, M., Cai, Q., Cheng, X., Shen, Y. & Liang, Z. (1995). Improved isolated microspore culture efficiency in medium with maltose and optimized growth regulator combination in japonica rice *(Oryza sativa). Plant Cell Tiss. Org. Cult.*, *42*, 245–250.

[250] Hoekstra, S., Van Bergen, S., Van Browershaveni, I. R., Schilperoort, R. A. & Heidekamp, F. (1996). The interaction of 2,4-D application and mannitol pretreatment in anther and microspore culture of *Hordeum vulgare* L. cv. Igri. *Plant Physiol.*, *148*, 696–700.

[251] Zheng, Y. & Konzak, C. F. (1999). Effect of 2, 4-dichlorophenoxyacetic acid on callus induction and plant regeneration in anther culture of wheat (*Triticum aestivum* L.). *Plant Cell Rep.*, *19*, 69–73.

[252] Shariatpanahi, M. E., Belogradova, K., Hessamvaziri, L., Heberle-Bors, E. & Touraev, A. (2006). Efficient embryogenesis and regeneration in freshly isolated and cultured wheat (*Triticum aestivum* L.) microspores without stress pretreatment. *Plant Cell Rep.*, *25*, 1294–1299.

[253] Reynolds, T. L. & Kitto, S. L. (1992). Identification of embryoid-abundant genes that are temporally expressed during pollen embryogenesis in wheat anther cultures. *Plant Physiol.*, *100*, 1744–1750.

In: Auxin
Editor: Rosario D. Pacovsky

ISBN: 978-1-61122-060-5

Chapter 5

INCLUSION COMPOUNDS OF AUXINS IN CYCLODEXTRINS

Kostas Bethanis*[*], Vasso Triantafyllopoulou and Dimitris Mentzafos

Physics Laboratory, Biotechnology Department,
Agricultural University of Athens, Athens, Greece

ABSTRACT

The low aqueous solubility of the auxins causes major problems in their administration. In order to overcome these problems water soluble salts of auxins are used or auxins are dissolved in ethyl alcohol (or another organic solvent). It has been shown that ethyl alcohol may have modifying effects in auxin action even at low concentration (Allerup, 1962). In addition, the alcohol evaporates during use, making the solution more and more concentrated and thus stunted or abnormal growth is observed.

An alternative way to increase the water solubility of the auxins is via their encapsulation inside suitable amphiphilic molecular capsules. Cyclodextrins (CDs) are oligosaccharides consisting of six, seven or eight α-(1→4)-linked D-glucopyranose units called α-, β- or γ-CD respectively, having a toroidal shape with a lipophilic inner cavity and a hydrophilic outer surface. As a result of their molecular structure and shape, they possess a unique ability to act as molecular containers by

[*] tel. ++302105294211, Fax: ++302105294233. E-mail: kbeth@aua.gr.

entrapping guest molecules in their hydrophophic cavity and thus exerting a profound effect on the physicochemical properties of guest molecules (Del Valle, 2004). Moreover, due to the weak non-covalent interactions between the host and the guest molecules, these supramolecular complexes are not fixed or permanent but rather in a dynamic equilibrium with their free components in solution, thus offering a slow release of the guests and sustaining their action.

CDs are used in agriculture, food, pharmaceuticals, cosmetics, environment protection and it is generally acceptable that the inclusion complex formation of auxins with CDs could improve their bioavailability, enhance their stability and reduce considerably the side effects following administration. The low cost, biocompatible and effective degradation makes *β*-CDs the best candidates for such molecular encapsulants. Commercial available derivatives of *β*-CDs are also very interesting because of their increased aqueous solubility and inclusion capacity.

In this chapter, studies on the molecular encapsulation of eight auxins (1-NAA, 2-NAA, 4-CPA, MCPA, 2-NOA, IBA, 2,4-D and 2,4,5-T) have been extensively reviewed. The molecules that have been used as hosts are mainly native *β*-CDs, methylated or hydroxypropyl *β*-CD derivatives and in a few cases *α*- or *γ*-CDs. The combined use of different analytical techniques has provided a complete characterization of the inclusion complexes. The structures of some of these complexes have been accurately determined in crystalline state by X-ray crystallography. NMR, spectrofluorometric and phase solubility studies have given solid evidences about the formation of the complexes in solution and estimations of the binding constants. DSC and FT-IR studies have checked the formation of a true solid inclusion compound and they have determined the stoichiometry of the formed complexes in the solid state.

This inclusion complexation could be regarded as an important step in the design of novel formulations of auxins with enhanced chemical properties and the gathered results can be regarded as an essential step to the development of controlled release agricultural formulations containing auxins. In addition, the inclusion of the auxins inside the CDs' hydrophobic cavity can be regarded as a simple prototype for ligand-enzyme systems, involving mainly van der Waals and hydrophobic interactions and therefore their structural study can provide some insight about the auxins binding into the pockets of the protein receptors.

ABBREVIATIONS

indole-3-butyric acid	IBA
4-chlorophenoxyacetic acid	4-CPA

2-methyl-4-chlorophenoxyacetic acid	MCPA
2,4-dichlorophenoxyacetic acid	2,4-D
2,4,5-trichlorophenoxyacetic acid	2,4,5-T
2-napthoxyacetic acid	2-NOA
2-naphthaleneacetic acid	2-NAA
1-naphthaleneacetic acid	1-NAA
α-, β-, γ- cyclodextrin:	α-, β-, γ-CD
hydroxylpropyl- β- cyclodextrin	HP-β-CD
methylated - β- cyclodextrin	Me-β-CD
2,6-di-O-methyl-β-cyclodextrin:	DIMEB or 14Me-β-CD
heptakis(2,3,6-tri-O-methyl)-β-cyclodextrin:	TRIMEB
cucurbit[7]uril	CB[7]
Physical Mixture	PM
Kneading method	KN
Co-evaporation method	CE
Spray-drying	SD

Graphical Abstract

IBA

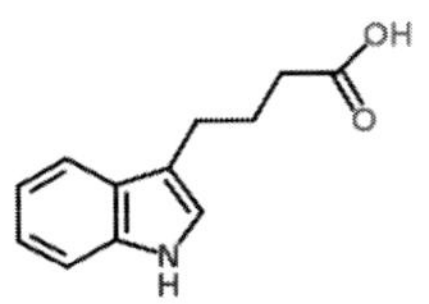

IBA/β-CD

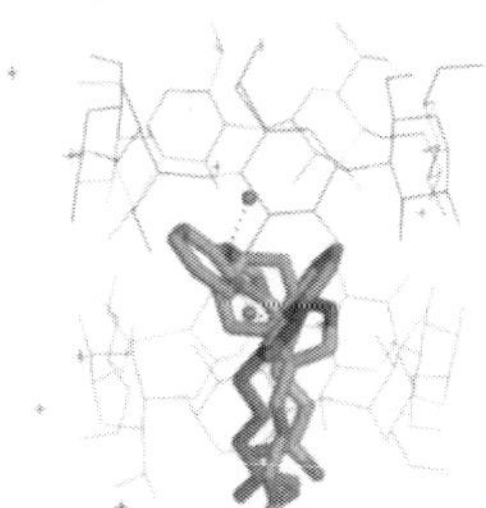

IBA/TRIMEB

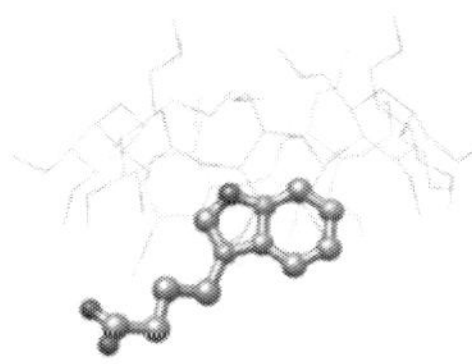

4CPA

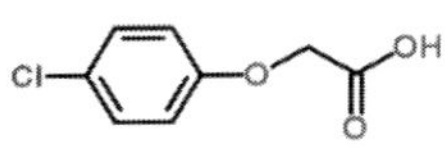

4-CPA /*β*-CD

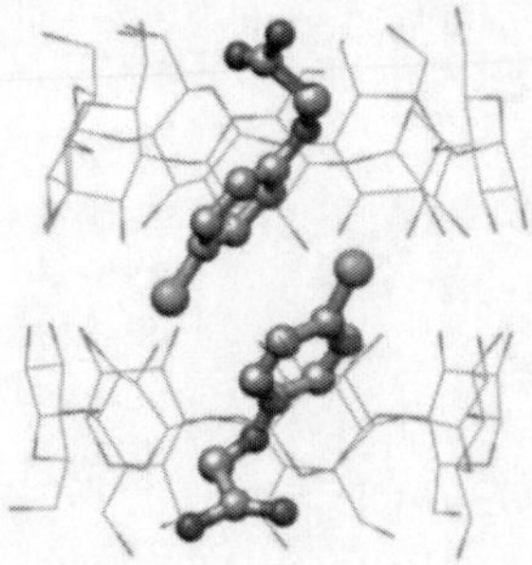

4-CPA/TRIMEB

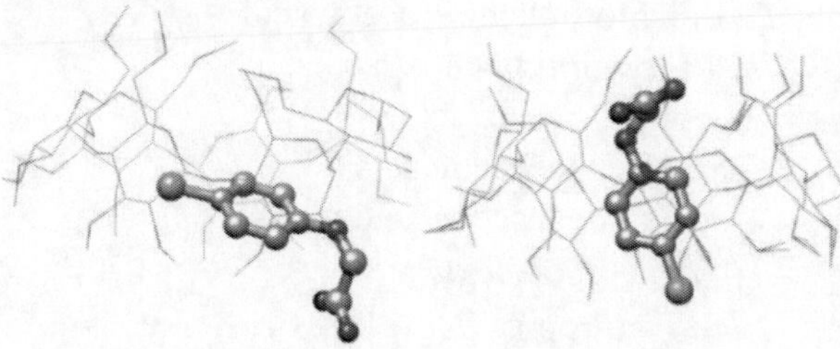

(a) (b)

MCPA

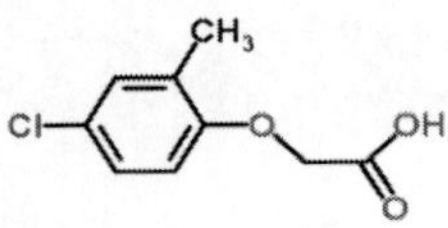

MCPA/*β*-CD

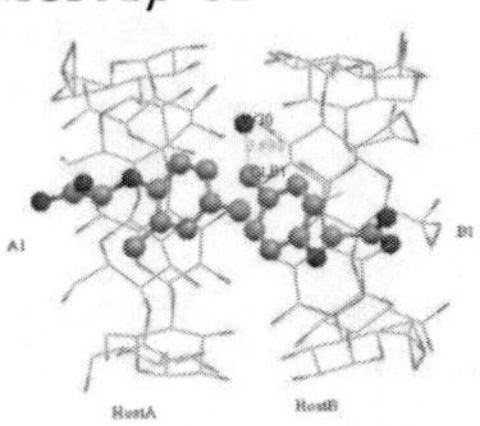

2,4-D

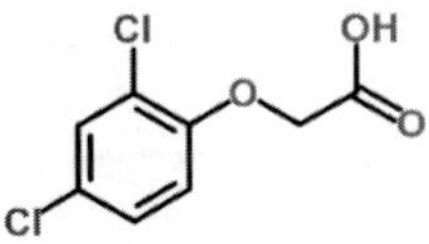

2,4-D/*β*-CD

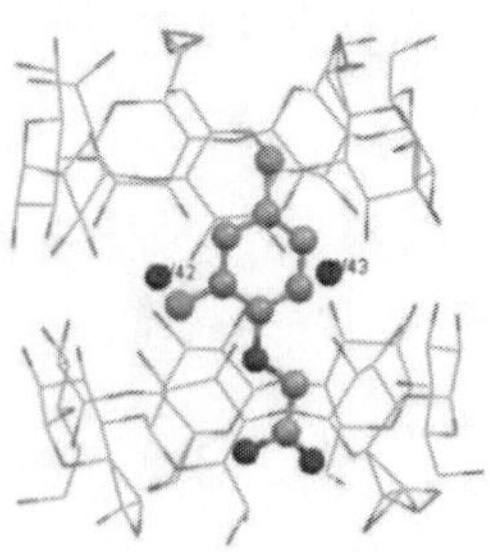

2,4-D/DIMEB

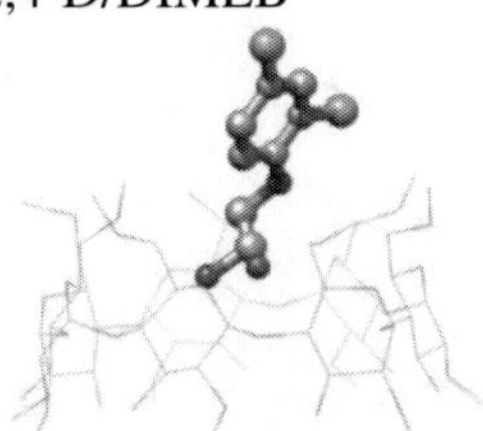

2,4-D/TRIMEB

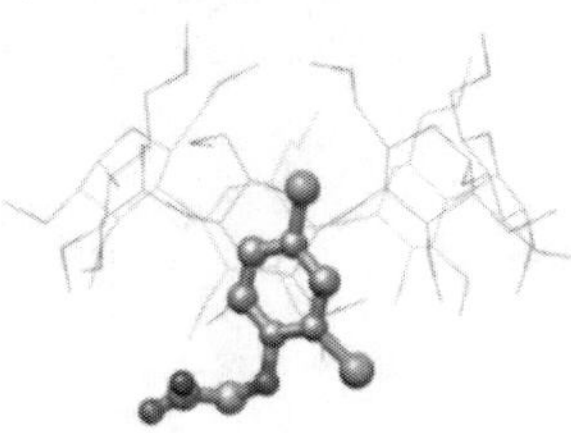

2,4,5-T

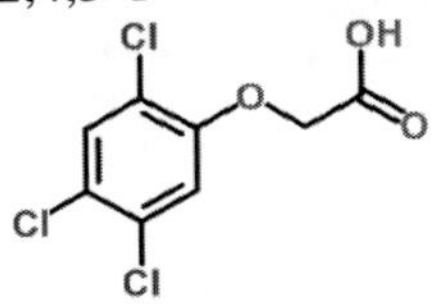

2,4,5-T/TRIMEB

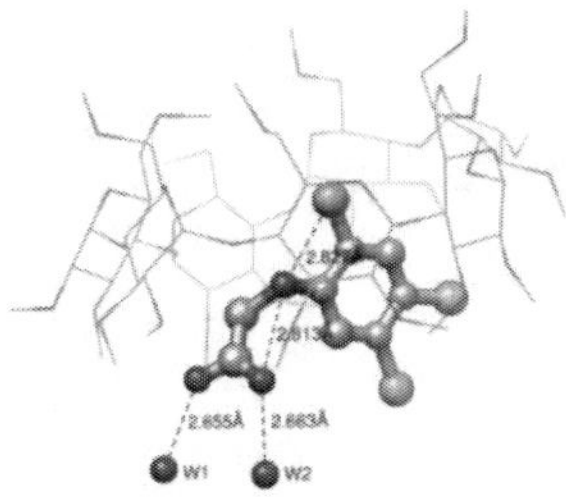

2-NOA

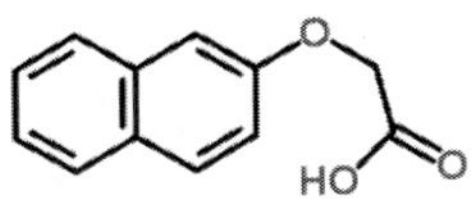

2-NOA/β-CD

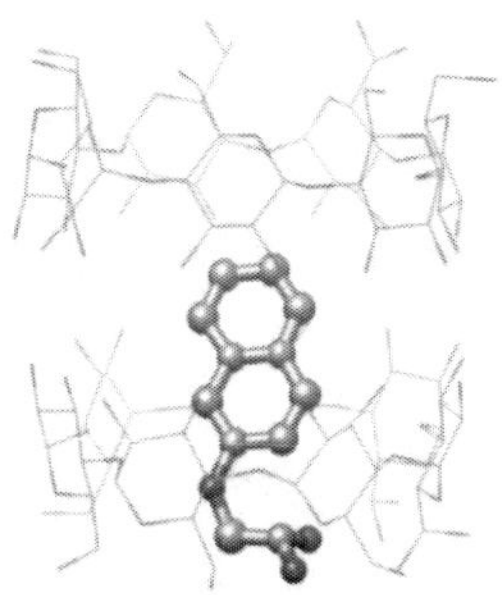

1-NAA

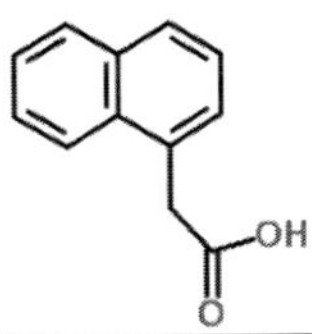

1-NAA/*β*-CD

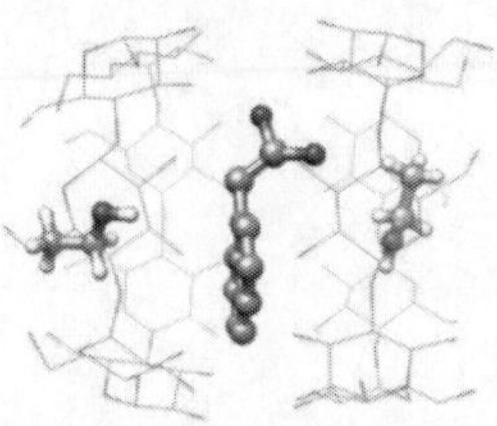

1-NAA/DIMEB

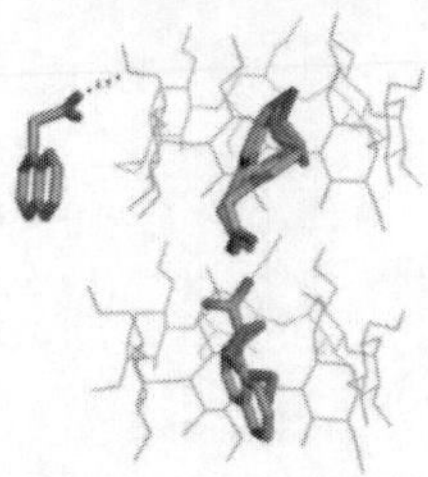

1-NAA/TRIMEB

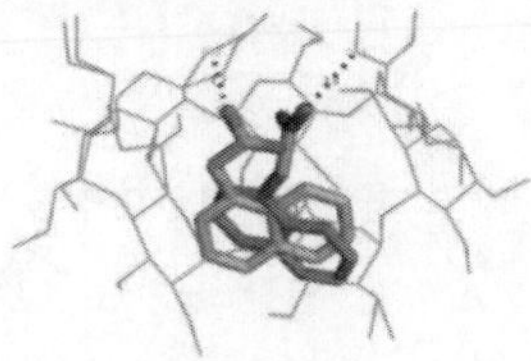

2-NAA

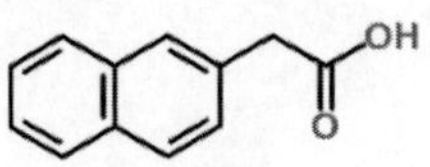

2-NAA/*β*-CD

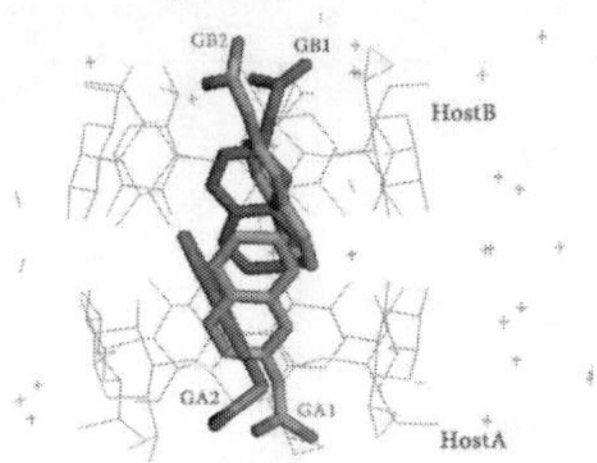

2-NAA/TRIMEB

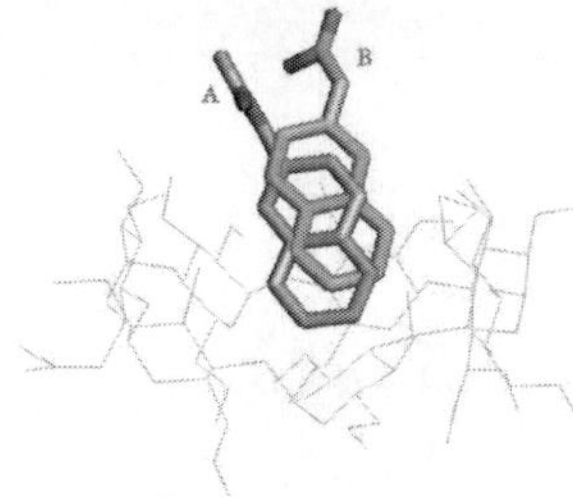

INTRODUCTION

Auxins are a class of plant growth hormones playing a key role in a wide variety of growth and developmental processes. At the cellular level, auxins act as signals for division, extension and differentiation during the course of the plant life cycle. At the whole-plant level, auxins play an important role in root formation, apical dominance, tropism and senescence (Guilfoyle et al., 1998). Except of the naturally occurring auxins more than 300 synthetic compounds are commercially available (Hence, 1990). However, the increasing use of some of them in agriculture causes environmental pollution (Hiller, 2008). Moreover, the auxins are characterized by a low aqueous solubility causing major problems in their administration. The common practice to overcome these problems is to dissolve auxins in an organic solvent like ethanol or use their water-soluble salts instead. Yet, it has been shown that ethanol may have modifying effects in auxin action even at low concentration (Allerup, 1962). In addition, the alcohol evaporates during use, making the solution more and more concentrated and thus leading to stunted or abnormal growth.

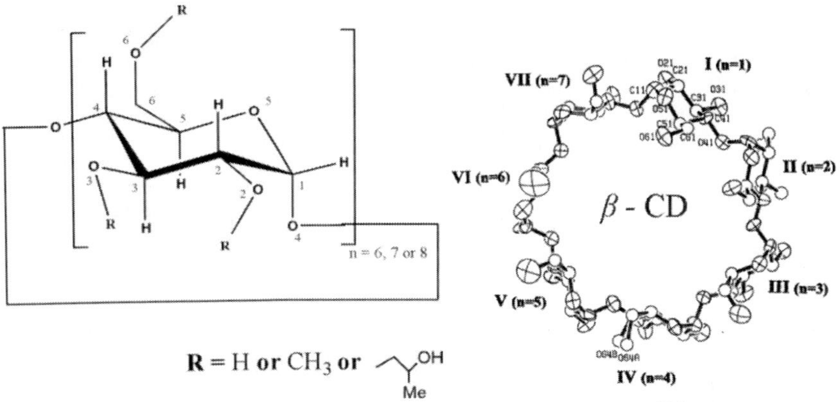

(a) *(b)*

Figure 1. (a) D-glucopyranoside unit (R = H, CH_3 or $-OCH_2CH(OH)CH_3$ for native CDs, methylated or hydroxypropyl CD derivatives respectively; n = 6, 7 or 8 for α-, β- or γ-CD respectively) (b) β-CD numbering scheme.

An alternative way to increase the water solubility of the auxins is via their encapsulation inside suitable amphiphilic molecular capsules. Cyclodextrins (CDs) are oligosaccharides consisting of six, seven or eight (n = 6, 7 or 8) *a*-(1→4)-linked D-glucopyranoside units, called α-, β- or γ-CD respectively (Figure 1). They are produced relatively simple from starch by means of enzymatic conversion using a set of easily available enzymes. The characteristic shape of CDs is that of an empty truncated cone with a lipophilic inner cavity and a hydrophilic outer surface (Figure 2). The narrower rim of the cone is the region of the primary hydroxyl groups of the CD (one in each D-glucopyranoside unit; O6*n*-H) whereas the broader rim is the region where the secondary hydroxyl groups of the CD (two in each D-glucopyranoside unit; O2*n*-H and O3*n*-H) are located.

Owing to their biological origin only dextrorotatory enantiomers of CDs are formed. Their glucopyranoside residues have the ${}^4C_1$ chair conformation, as it is indicated by the Cremer-Pople Q and θ puckering parameters (Cremer, 1975). The primary hydroxyl groups, O6*n*-H, can freely rotate and they have usually the *gauche-gauche* orientation pointing outwards the CD cavity. Some of them may have the *gauche-trans* orientation pointing inwards the cavity, blocking it partially. The *trans-gauche* orientation is unfavorable energetically and thus extremely rare. It has been observed only in a few cases of substituted CDs (Guo, 2011; Anibarro, 2001).

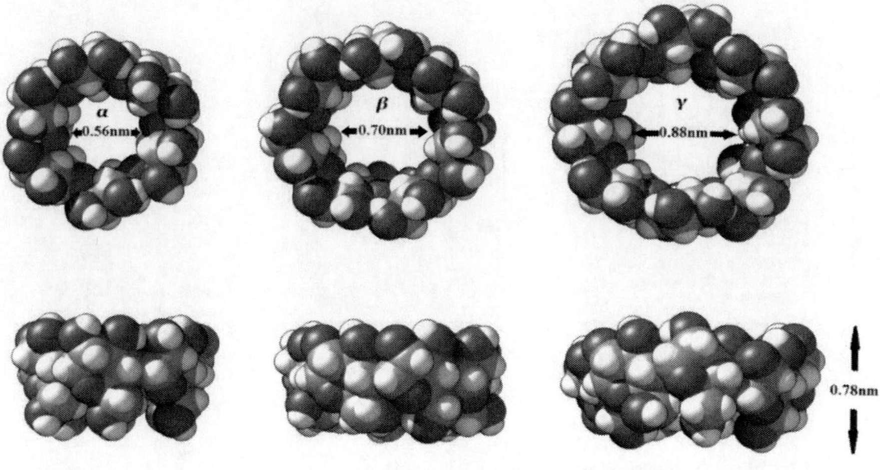

Figure 2. 3D shape of the *α*-, *β*- and *γ*-CD and dimensions of their cavity.

Unlike the primary hydroxyl groups which have the ability to rotate, the secondary hydroxyl groups of the native CDs (and in some cases of their derivatives) participate in a hydrogen bond intramolecule network (O3*n*-H of the nth glucosidic residue is hydrogen bonded to the O2(*n*+1)-H of the adjacent $(n+1)^{th}$ glucosidic residue) giving rise to a rigid form to the CD truncated cone (Uekama, 1998).

As a result of their molecular structure and shape, the CDs possess a unique ability to act as molecular containers by entrapping guest molecules of suitable size inside their hydrophobic cavity. This encapsulation exerts a profound effect on the physicochemical properties of the guest molecules (Del Valle, 2004). Moreover, due to the nature of the weak non-covalent interactions between the host and the guest molecules, these supramolecular complexes are not fixed or permanent but rather in a dynamic equilibrium with their free components in solution, offering a slow release of the guests and sustaining their action. Several factors contribute to the inclusion of guest molecules inside the CD cavity such as hydrogen bonds, the hydrophobic effect, expulsion of water molecules, release of strain energy in the CD ring and polar interactions between host and guest molecules (Mentzafos, 1991).

For all the above mentioned reasons, CDs are widely used in agriculture, food, pharmaceuticals, cosmetics, environment protection and it is generally acceptable that the formation of the inclusion compounds of auxins with CDs could improve their bioavailability, enhance their stability, reduce considerably the side effects following administration and sustain their activity for a longer time period due to their slow release from the CD cavity. Furthermore, CDs act as precursor solubilizers, playing probably a role of protecting agents and reactant carriers as their concentration remains unchanged during product formation by the plant cells (van Uden, 1994). Therefore, the inclusion complexation could be regarded as an important step in the design of novel formulations of auxins with enhanced chemical properties and in the development of controlled release agricultural formulations.

In designing auxin/CD inclusion compounds, the choice of the appropriate host molecule, among the native and modified CDs, depends mainly on the size of the guest molecule and the stability of the complex that one wishes to achieve. Of all the common native CDs (*α*-, *β*- or *γ*-CD), *β*-CD has been studied more extensively because of its cavity size, its low cost, its biocompatibility and effective degradation. The native CDs can be chemically modified by hydroxyalkylation, alkylation, sulfoalkylation, etc. The principal aim of such modification is to increase the solubility and safety profile of the parent native cyclodextrins. Commercially available methylated or hydroxypropyl *β*-CD derivatives have attracted interest because of their larger and more flexible cavity. The flexibility of this kind of CD derivatives is a result of the substitution of the CD hydroxyls and thus the lack of the intramolecular hydrogen bonds formed regularly in the native CDs between the adjacent D-glucopyranoside units. This leads to a significant increase of the CD inclusion capacity and aqueous solubility which allows extension of their applications.

In this chapter we present a review along with some new results on the formation, characterization and structure determination of the inclusion compounds of several auxins (endogenous and synthetic) with various native and derivative CDs. In particular, the inclusion complexes of the following auxins are examined:

1-naphthaleneacetic acid (1-NAA), 2-naphthaleneacetic acid (2-NAA),
4-chlorophenoxyacetic acid (4-CPA), 2-methyl-4-chlorophenoxyacetic acid (MCPA),
2-naphtyloxyacetic acid (2-NOA), indole-3-butyric acid (IBA), 2,4-dichlorophenoxyacetic acid (2,4-D) and 2,4,5-trichlorophenoxyacetic acid (2,4,5-T).

The chosen host molecules are: native α-, β- and γ-CD; hydroxylpropyl-β-CD (HP-β-CD); methylated-β-CD (Me-β-CD); 2,6-di-O-methyl-β-CD (DIMEB) and heptakis(2,3,6-tri-O-methyl)-β-CD (TRIMEB).

This inclusion complexation could be regarded as an important step in the design of novel formulations of auxins with enhanced chemical properties. The gathered results can be regarded as an essential step to the development of controlled release agricultural formulations containing auxins.

The effect of cyclodextrins (CDs) on the rooting promotion by auxins has been studied in vivo as well as in vitro in the cases of olive cuttings (Mura et al., 1995) and *Cynara scolymus cv. Early French* (Brutti, et al., 2000). These experiments have shown that the incorporation of α- or β-CDs to the rhizogenesis medium supplemented with 3 mg l^{-1} naphthalene acetic acid, allowed tripling the rooting percentage and doubling the number of differentiated roots per explant compared to medium without cyclodextrins. Mura et al. (1995) attributed the promotion of rhizogenesis to the increase in solubility of the auxin employed. However, the in vitro increase in rooted sprouts of the *cv. Early French* artichoke cannot be explained in this way, since the auxin is soluble at the concentration used. As CD increases cell-membrane permeability (Uekama et al., 1998) it may be speculated that altered interaction between auxins and their receptors, could explain the enhancing effects of CD on rhizogenesis of *Cynara scolymus cv. Early French*. In any case, the application of CD seems useful for the micropropagation of diverse species having horticultural interest and presenting rooting difficulties.

Moreover, a simple and efficient method for separation and determination of auxins by capillary electrophoresis has been developed based on the encapsulation of auxins in CDs. In particular, Liu et al. have successfully demonstrated a suitable method for the determination of indole acetic acid (IAA) in a real plant extract sample (Liu et al., 1996). Among the three native CDs, α- and γ-CDs provided better resolution than β-CD. This is probably due to the low aqueous solubility of the β-CD causing a less significant effect on migration behavior. Optimum separation was obtained with modifier systems consisting of mixtures of the CDs.

Finally, the inclusion of the auxins inside the CDs hydrophobic cavity can be regarded as a simple prototype for ligand-enzyme systems, involving mainly van der Waals and hydrophobic interactions and therefore their structural study can provide some insight about the auxins binding into the pockets of the protein receptors such as TIR and ABP1 (Triantafyllopoulou, 2013).

The molecular encapsulation of auxins inside a macrocyclic host other than CDs has also been investigated in a recent paper (Nuzzo et al., 2014). In this work the inclusion of the bioactive auxin molecules IAA, 2-NAA, and 2,4-D within the cucurbit [7] uril (CB [7]) and the influence of pH on the formation of these host-guest complexes, was studied by ^{1}H NMR and relaxation experiments. Cucurbit[*n*]urils (CB[*n*]) are an interesting family of host molecules structurally and functionally different from CDs. Cucurbit [7] uril was proved capable of hosting the protonated forms of the investigated plant growth molecules within its hydrophobic cavity, whereas the anionic forms of the auxins are released from the macrocycle thereby suggesting pH control over the sequestration and release of auxin molecules within CB [7]. Such a system has the advantage of being readily triggered at any desiderable time by simply and reproducibly adjusting the soil medium pH.

Materials and Methods

Chemicals

For the formation of the studied complexes, the auxins, the cyclodextrins and all the other reagents and solvents were purchased from companies of fine chemicals. Deionised water was used throughout all the experiments.

Preparation of the Complexes

Various methods are used and widely reported for the formation of the cyclodextrin inclusion complexes. The selection of the method and its effectiveness highly depends on the nature of the drug and/or cyclodextrin (Hedges, 1998; Challa et al., 2005). Auxin/CD solid systems are performed using different techniques, namely by kneading and co-evaporation. To perform comparative evaluations, with the respective solid inclusion complexes, physical mixtures are also prepared. In order to conduct X-ray crystallography experiments and determine the complex structure in the crystalline state, crystallization processes for single crystal growth of the complex are followed.

Physical Mixtures (PM)

Weighted amounts of auxin and cyclodextrin are simply mixed in a mortar during a time period of about 15 min.

Kneading Method (KN)

Auxin and cyclodextrin are accurately weighed and mixed together in different stoichiometric ratios (e.g. 1:1, 2:1, 3:1 and 1:2) using a mortar. Each mixture is then triturated and an appropriate amount of ethanol is added until a homogenous paste is formed. The paste is then kneaded for about 45 min and dried at 45 ◦C in an oven. The dried product is ground powdered and sieved.

Co-Evaporation Method (CE)

Auxin/CD complexes in a 1:1 molar ratio are prepared using the co-evaporation method. Auxin and CD (equimolar ratio) are completely dissolved in a solution of ethanol and water (v/v = 1:20). The dispersion of auxin in the aqueous CD solution is mechanically shaken at room temperature and 100 rpm for a time period varying from 24h to 4 days to achieve equilibrium of the complexation reaction. After evaporation of the ethanol from the reaction mixture, the uncomplexed auxin is removed by filtration. The filtrate is evaporated under reduced pressure to remove the solvent and dried in vacuum to give the auxin/CD complex.

Spray-Drying (SD)

Spray-drying is carried out in a Spray-Dryer equipment. For this purpose, auxin is dissolved in an organic solvent (e.g. ethanol) and an amount of cyclodextrin (auxin:cyclodextrin molar ratios 1:1 or 1:2) is separately dissolved in purified water. The solutions are next mixed for about 20 min by sonication to produce a clear solution which is spray-dried under specific conditions.

Crystallization

The auxin is mixed in an aqueous solution of native or methylated β-CD (equimolar ratio) and the mixture is stirred for a two hours period. In some cases ethanol is added (ethanol/water v/v = 1:20) in order to dissolve the insoluble auxins. Only in the case of the crystal structure of the 1-naphthylacetic acid/β-CD complex determined by Wang et al. (Wang, 2011) the solvent used was ethanol:water v/v = 1:2 affecting significantly the formation of the inclusion complex as described in a next paragraph. In any case the resulting precipitate of the β-CD complex is redissolved at 65-70 °C and left to cool to room temperature over a period of one or two weeks. The solutions of auxins with DIMEB or TRIMEB hosts are stirred for 30 min at 45°C and then maintained at 50°C for some days. At the end of these time periods transparent crystals of the complexes, suitable for X-ray data collection are obtained.

Methods

The recent trend in CD complex studies is the combination of several experimental techniques, which include X-ray crystallography, various spectroscopic methods, such as FTIR, UV-Vis, circular dichroism, ^{1}H and ^{13}C NMR spectroscopy, microcalorimetry (DSC and ITC) and molecular modeling. A brief description of the techniques that have been applied so far for the study of some auxin/CD complexes is following whereas the reported results are reviewed more analytically in the next paragraphs.

UV-Visible Spectroscopy

Spectrophotometric measurements are performed to quantify auxins in their free and CD-complexed form. Standard curves of the auxins are prepared in deionised water. Spectrophotometric scans are performed between 190 and 300 nm to monitor the UV spectra of the auxins. The wavelength of the maximum absorbance is then used to quantify auxin concentration.

Spectrofluorometric Studies

In luminescence studies, CDs have been employed to enhance fluorescence emission of different luminophors and to induce room temperature phosphorescence under appropriate conditions (Munoz de la Pena et al., 1993). The intensification of luminescent processes of molecules included in the interior of the cavity of the CD is due to the better protection from quenching and other processes occurring in the bulk solvent. CDs have been used to increase the fluorescence intensity of various organic species through a partial encapsulation or total inclusion. Solutions are excited at the wavelengths of maximum absorbance of the examined auxin complex (λ_{max}) and fluorescence measurements are made with particular excitation and emission band widths. Absorption measurements are conducted with a spectrophotometer connected to a computer and specific software is used for spectral acquisition and analysis of the spectrophotometric data.

Phase Solubility Studies

Phase solubility studies are carried out according to the method described by Higuchi and Connors (Higuchi & Connors, 1965). Excess amount of an auxin is added to aqueous solutions containing various concentrations of the host cyclodextrin. Then, the suspensions are shaken on a rotary shaker at room temperature for various periods of time (a few hours to 4 days). After equilibrium is reached, suspensions are centrifuged and the supernatant is withdrawn and properly diluted. The concentration of the auxin is determined by spectrophotometry (UV–Vis Spectrophotometer) at the wavelength at which the auxin has the highest absorption (standard curves of each auxins were prepared in deionized water. The apparent stability constants Ks are calculated from phase solubility diagrams with the assumption of 1:1 stoichiometry according to equation:

$$Ks = \text{slope} / S_0(1 - \text{slope})$$

where S_0 is the solubility of the auxin in absence of CDs. The assumption of 1:1 stoichiometry is based on the type of the diagram obtained. In the reported auxin/CD phase solubility studies, stability constants of the complexes were estimated only for the diagrams that were classified as of A_L type according to Higuchi and Connors (Higuchi & Connors, 1965) and thus correspond to a 1:1 complex stoichiometry.

Infrared Spectroscopy (FTIR)

FTIR spectroscopy is a technique that can provide information about the chemical bonds and molecular structure of the "guest" molecules. For this purpose the FTIR spectra of the auxin, the cyclodextrin, the physical mixture and the inclusion complexes are recorded at room temperature in a spectral region between 4000 and 400 cm^{-1}.

Thermal Analysis

Two microcalorimetric methods have found widespread application in CD studies and drug-design research. These two methods are the *Differential Scanning Calorimetry* (DSC) and the *Isothermal Titration microCalorimetry* (ITC). DSC and ITC can be considered as complementary tools for investigating molecular recognition between a host and guest in both solution and the solid state.

In DSC, the sample under investigation (liquid or solid) is gradually heated/cooled at a certain speed and the molecular events caused by the temperature change (phase transitions, melting, decomposition, etc.) are recorded as heat release or absorption. When guest molecules are embedded into β-CD cavities, their melting, boiling or sublimating points, generally shift to a different temperature or disappear. In accordance, the DSC profiles of pure components (auxin, cyclodextrin) and binary systems are acquired.

In ITC, one component of the relevant species involved in the reaction (association, aggregation, or molecular recognition) is introduced in a stepwise fashion into a solution of the other component(s), with all the molecular events occurring upon this addition registered in a thermogram.

By simultaneously using both methods, we can obtain complete thermodynamic information about the system under investigation at various molar ratios as well as at different temperatures (Rekharsky and Inoue, 2006).

Nuclear Magnetic Resonance (NMR)

Nuclear magnetic resonance (NMR) spectroscopy is one of the most powerful and versatile methods for the elucidation of molecular structure and dynamics. It is also very well suited to study molecular complexes and their properties. Therefore, it has been widely used for studying inclusion complexes formed by cyclodextrins. The success of NMR spectroscopy in this field is due to its ability to study complex chemical systems and to determine stoichiometry, association constants and conformations of molecular complexes, as well as to provide information on their symmetry and dynamics. Furthermore, compared to other techniques, NMR spectroscopy provides a superior method to study complexation phenomena, because guest and host molecules are simultaneously observed at the atomic level (Ejchart and Kozminski, 2006).

In Silico Studies

The main computational calculations of CD complexes modeling concern Quantum Chemical Calculations (QC), Molecular Mechanics (MM) and Molecular Dynamic Simulations (MD).

MM simulations are performed in order to model the effect of the pattern of substitution (functional groups) in the molecular recognition between the auxins and the different CDs under study. The software used for this purpose in the studies presented in this chapter is the Tinker software suite (Ponder et al., 1987) with the Optimized Potentials for Liquid Simulations (OPLS) (Jorgensen et al., 1996) combined with the TIP3P model for explicit water (Jorgensen et al., 1983). OPLS was chosen as it includes parameters specifically derived for carbohydrates. Implicit solvent calculations were computed with the Generalized Born model augmented with an accessible Surface Area term (GBSA) (Qiu et al., 1997; Hou et al., 2011).

MD simulations of the auxin/CD complexes were performed with the *Desmond* program (Shivakumar et al., 2010) from the Schrodinger suite. The OPLS2005 force field (Banks et al., 2005) was applied for the simulations. Before simulation the complex was solvated with TIP3P waters (Mark & Nilsson, 2001) to form a water shell 10 Å thick around the complex. The solvated host guest system was simulated for a period of 10 ns with the 'NPT relax protocol' in Desmond. The protocol involves an initial minimization of the solvent with the solute restrained. The minimization was followed by short MD simulations of 12 - 24 ps in sequential NVT and NPT ensembles with the Langevin thermostat and barostat (Quigley & Probert, 2005). The temperature was maintained by coupling to an external 300 K bath based on the Langevin algorithm. The pressure was isotropically restrained to 1 bar with the Langevin barostat. High-frequency vibrations were removed by applying the SHAKE algorithm (Ryckaert et al., 1977). Initial velocities are generated randomly from a Maxwell distribution at 300 K in accordance with the masses assigned to the atoms. After completion of the relaxation protocol, an unrestrained 5-ns MD simulation was carried out on the solvated complex, avoiding introduction of any structural artifacts. During the simulation, the geometries and the corresponding energies of the complex were sampled every 5 ps. After the trajectories were obtained from the MD simulations, the structural stability of the complex was analyzed based on the root-mean-square deviation (RMSD) computed using the (first frame) starting structure as a reference. On completion of the simulation, the binding energies and intermolecular hydrogen bonds were computed with the Simulation Event Analysis module in Desmond.

However, it should be stated that the results of a modeling of a CD complex should not be presented as "hard" findings without confronting them with experimental data. These results can reveal some trends rather than ''absolute'' values and one should be very cautious in order to avoid misinterpretation of the experimental data. Therefore, until a comprehensive evaluation of theoretical approaches is carried out, CD complexes modeling can, at best, provide purely qualitative results (Dodziuk, 2006).

X-Ray Crystallography

Diffraction Data Collection

The data sets were collected using *in house* Rigaku X-ray rotating anode and a Syntex $P2_1$ diffractometer source or synchrotron radiation. Crystals of good quality were selected and enclosed in Lindermann glass capillaries to prevent loss of water and radiated by a Ni-filtered Cu Kα X-ray beam (λ = 1.5418 Å). A poor diffracting power appeared in all the crystal complexes due to the variability and disorder of the water contents and of the guest molecules. The intensities were corrected for Lp and absorption effects by the Pi-scan process. In the case of the IBA/*β*-CD complex, the data set was collected on a D8 VENTURE system equipped with a multilayer monochromator and a Cu Incoatec microfocus sealed tube (λ = 1.54178 Å) and the frames were integrated with the Bruker *SAINT* Software Package using a narrow-frame algorithm.

Low-temperature and high resolution data of the complexes were collected using the synchrotron radiation light source at an EMBL beamline at the DORIS storage ring, DESY, Hamburg, using a marCCD 165 detector. The crystals were picked with a cryoloop and flash cooled under the N_2 stream to 100 K. Parafine oil was used as a cryoprotectant of the crystals. In the cases where the cryoprotection was not adequate and the crystals were damaged by the shock cooling under the cryo-conditions, the data set was collected at room temperature. In some cases, in order to have a complete data set, data collection was performed twice, for high and low resolution. The programs *DENZO* and *SCALEPACK* (Otwinowski, 1997) or *KABSCH* (Kabsch, 1993) were used for data processing and scaling.

Crystal Structure Determination

The structures have been solved by a Paterson-vector search method and Fourier recycling using the program *DIRDIF2008* (Berskens, 2008) with the coordinates of the macrocycle of a known complex as an original model. In the cases where an isostructural complex was known the new complex structure was just refined based on F^2 using the *SHELXL97* program (Sheldrick, 2008). In any case, subsequent difference electron maps ($\Delta\rho$) revealed the positions of the remaining non-H atoms of the host, the guest and the water molecules. The used programs and the graphical display were parts of the *WINGX* suite (Farrugia, 1999). When the number of the observed data reflections of the complexes was not sufficient it was possible to use anisotropic thermal parameters only for selected non-hydrogen atoms. Hydrogen atom positions linked to primary, secondary or tertiary carbon atoms of the host molecules of the structures were imposed computationally at calculated C-H distances and their thermal parameters were set to 1.2 U_{iso} of the isotropic thermal parameter of the corresponding carbon atom.

Data collected with the Syntex P2_1 diffractometer gave relatively high *R* factors a characteristic frequently observed in the structures of the CD complexes because of the high disorder of the guest, the water molecules and some of the hydroxyl or methoxy groups of the hosts. As a result the number of the observed data was limited not allowing the anisotropic treatment of all the non-H atoms. In some cases, like the highly disorder IBA/β-CD structure, the failure of structure refinement with all the conventional crystallographic programs, led us to the development of novel methods and algorithms for crystal structure determination like the *TWIN* algorithm (Bethanis et al., 2000; 2000a, 2007).

INCLUSION COMPOUNDS OF AUXINS WITH CDS

Indole-3-Butyric acid (IBA)

The IBA/β-CD inclusion compound has been prepared and studied with different spectroscopic methods. The inclusion of IBA in β-CD has been studied in solution by the phase solubility diagram which has given an estimation of the stability constant (Ks) of the complex and in the solid state by FT-IR spectroscopy (Fourtaka, 2012). Moreover, the crystal structure of the complex has been determined by X-ray crystallography (Mazi, 2012). Finally, molecular dynamics simulations of the complex have been performed (unpublished data) in order to gain an insight of its dynamic behavior in solution.

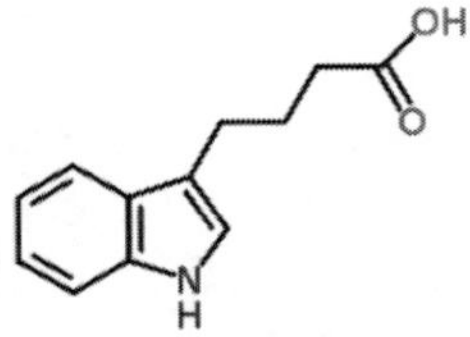

Figure 3. Scheme of the Indole-3-butyric acid (IBA) molecule.

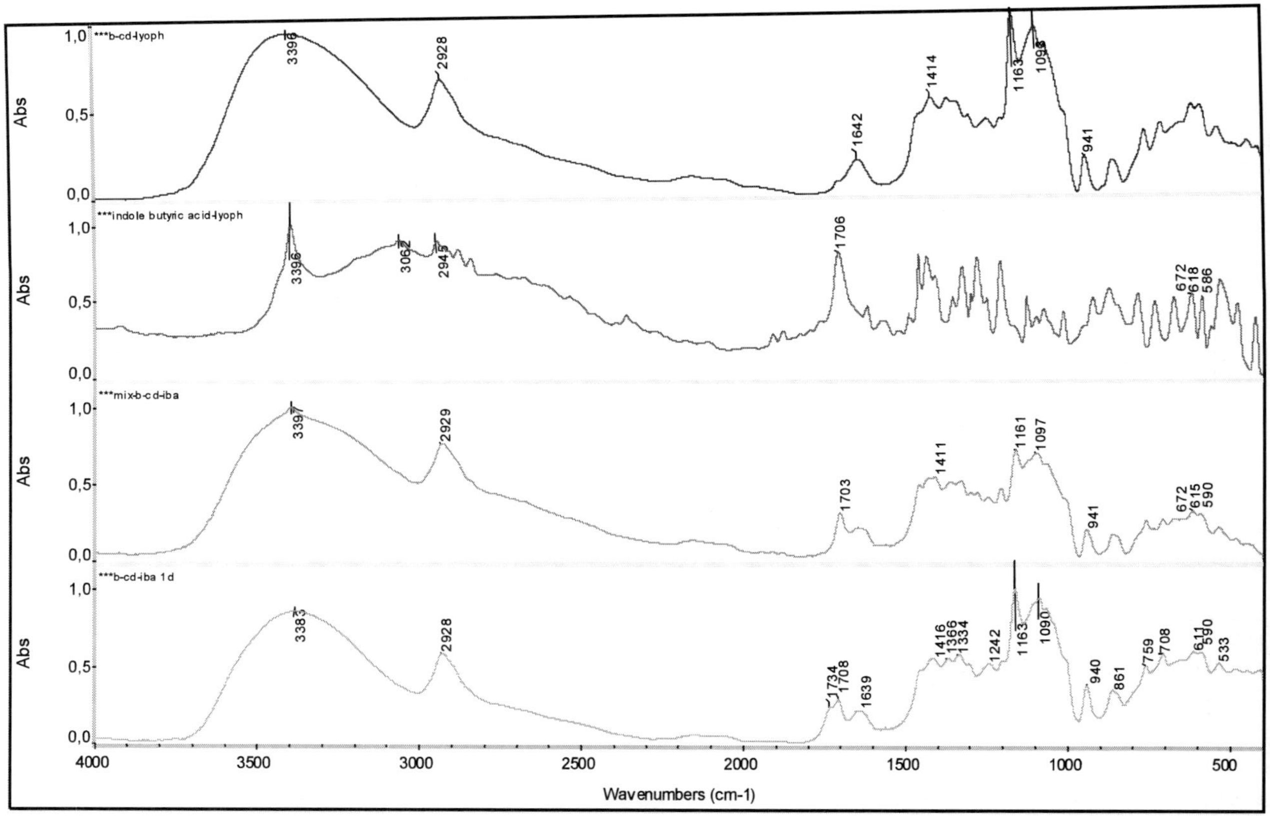

Figure 4. FT-IR spectra of β-CD (blue), IBA (red), physical mixture (green) and inclusion complex (cyan).

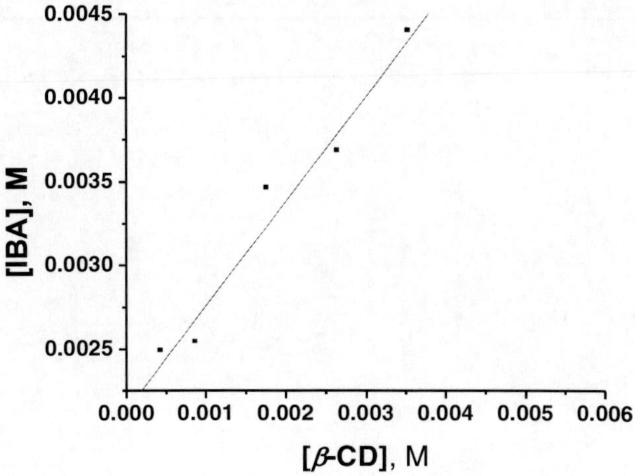

Figure 5. Phase solubility diagram obtained for IBA and increasing concentrations of *β*-CD in water at 25°C.

Infrared Spectroscopy (FT-IR)

Figure 4 presents the infrared spectra of *β*-CD, IBA, physical mixture and inclusion complex. The FT-IR spectrum of IBA shows a characteristic absorption band at 1642 cm^{-1} assigned to carboxylic C=O stretching vibration and a wide band due to hydroxyl O-H of the carboxylic acid group that appears at 3396 cm^{-1}.

In the physical mixture, no variations on these wavenumber values were observed. Nevertheless, the spectra of the inclusion complex showed a displacement of the band corresponding to C=O stretching vibration, 1703 and 1734 cm^{-1} respectively. The overall FT-IR data highlight the formation of an inclusion complex in which IBA is probably located inside the *β*-CD cavity. In addition significant intermolecular hydrogen bonding interactions were detected between the butyric acid side chain of the auxin and the CD. In fact, the FT-IR spectra reveal important changes in some vibrational modes as consequence of the host/guest interactions.

Table 1. Complex formation parameters for IBA /*β*-CD system in water at 25°C

S_0	α	Ks (M^{-1})	R^2
0.002	0.629	792	0.982

S_0: represents the solubility of IBA, α: slope, Ks: stability constant.

Phase Solubility Studies

The stability constant of IBA/β-CD was derived from the changes in the solubility of IBA in the presence of increasing amounts of β-CD, measured by UV spectrophotometry. Figure 5 presents the phase solubility diagram obtained from IBA and increasing concentrations of β-CD in water at 25°C. The diagram indicates that IBA solubility is increased by the presence of the macrocycle and it is classified as of A_L type according to Higuchi and Connors (Higuchi & Connors, 1965) as the IBA solubility increases linearly with increasing β-CD concentrations. The stability constant, Ks, was therefore calculated from the straight line of the phase solubility diagram according to the equation previously presented. Table 1 summarizes data obtained from the phase solubility diagram (IBA solubility, slope, stability constant and correlation coefficient).

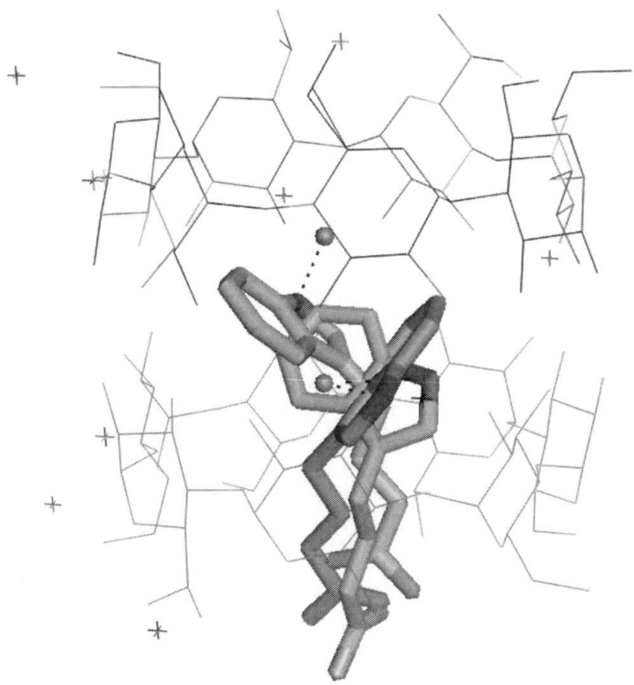

Figure 6. Crystal structure of the encapsulated IBA inside the formed dimeric cavity of a β-CD dimer. IBA has been found occupying three possible sites denoted with different colors (cyan: site A with occupancy factor (o.f.) = 0.4, magenta: site B with o. f. = 0.3 and green: site C with o.f. = 0.3).

IBA /β-CD Crystal Structure

The crystal structure determination of the IBA/β-CD complex has revealed the following (Mazi, 2012): one IBA molecule has been found encapsulated inside the cavity formed by a head to head β-CD dimer. Therefore, the host:guest stoichiometry is 2:1. The guest molecule has been found disordered over three sites, A, B and C (Figure 6). The complex crystallizes in Space group *C*2 and the molecular packing is that of a channel (*CH*) (Mentzafos et al., 1991).

A Occupied Site

The indole plane is inclined by 38.2° with respect to the O4*n* mean plane and the butyric acid is located at the primary hydroxyl groups region. One oxygen atom of the carboxyl group of the guest is hydrogen bonded with its homologue of the vicinal dimer of the channel. The other oxygen atom of the carboxyl group is hydrogen bonded with a hydroxyl group of its own host β-CD.

B Occupied Site

The indole plane is inclined by 60° with respect to the O4*n* mean plane. The carboxyl group protrudes from the cavity. The carboxyl atom of the guest is hydrogen bonded with the hydroxyl group of a vicinal host molecule.

C Occupied Site

The indole ring is located at the intradimeric region of the dimer forming an angle of 49.4° with the O4n mean plane of the O4*n* glucosidic atoms. The carboxyl group protrudes from the primary hydroxyl groups region being hydrogen bonded with the primary hydroxyl of the host of a vicinal dimer.

In Silico Studies: Molecular Dynamics Simulations of IBA/β-CD Complex

Starting from the crystallographic coordinates of the IBA/β-CD complex corresponding to a formed β-CD dimer and one encapsulated IBA molecule, three molecular dynamics simulations, in acidic (pH = 1), neutral (pH = 7) and alkaline (pH = 12) solution were performed and analyzed (Fourtaka et al., unpublished results). The coordinates were subjected to equilibration and subsequent molecular dynamics simulations at 300 K for 10 ns. In all the cases, the initial β-CD dimer was decomposed after about 2ns. That means that monomer IBA/β-CD complexes are more possible to exist in the solution whereas the β-CD dimers appear in the crystalline state due to the crystal packing. The IBA molecule is always accommodated inside the cavity of the one β-CD molecule whereas several water molecules are found to be hosted inside the other CD cavity. In Figure 7 it is shown the time evolution of the root mean square deviation (RMSD) from the initial structure, calculated for all the IBA atoms. In acidic solution (pH=1) the behavior of IBA resembles that of IBA in the neutral pH solution. However, the RMSD value of IBA fluctuates much more significantly in the basic aqueous solution probably indicating a more unstable complex.

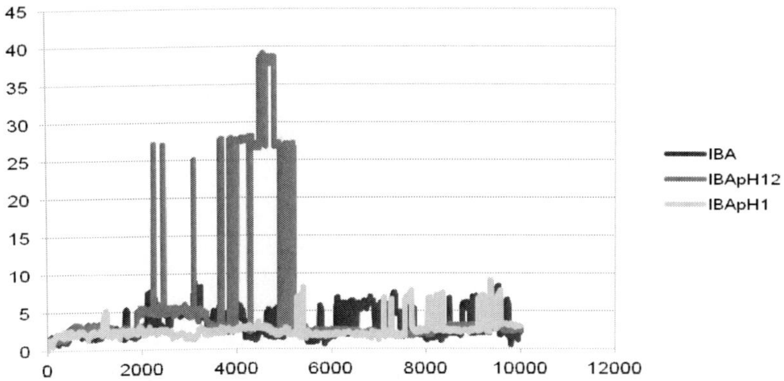

Figure 7. Time evolution (ps) of the RMSD (Å) from the initial structure, calculated for all the IBA atoms in acidic (pH = 1), neutral (pH = 7) and alkaline (pH = 12) solution.

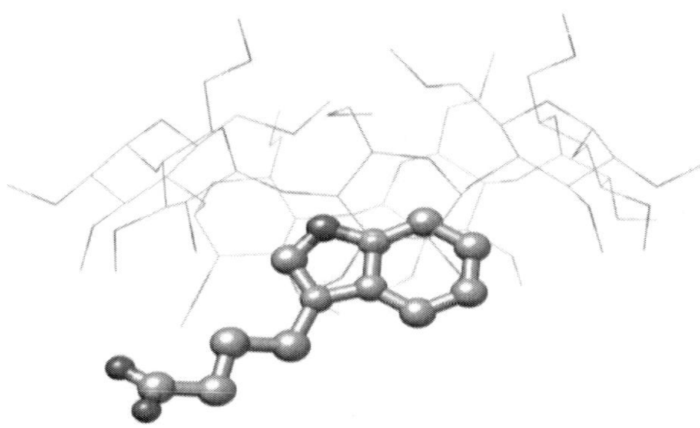

Figure 8. Crystal structure of IBA/TRIMEB complex. The primary rim of TRIMEB is noticeably narrower than that of the native β-CD forcing IBA to have an 'equatorial' orientation compared to the 'axial' orientation observed in the IBA/β-CD complex.

IBA/TRIMEB Crystal Structure

The crystal structure determination of the IBA/TRIMEB complex (Tsorteki et al., 2004) has shown that the complex crystallizes in the space group $P2_1$ stacking along the shortest a-axis. One water atom is entrapped inside the cavity having a low occupation factor. The indole moiety of the guest is accommodated at the secondary methoxy groups region in an 'equatorial' mode being almost perpendicular to the mean plane of the O4n glucosidic atoms (Figure 8). The butyric chain of the guest protrudes from the host cavity and its carboxyl group is hydrogen bonded with an oxygen atom of the secondary methoxy group of the host molecule of a vicinal complex. The nitrogen atom of the guest is also hydrogen bonded with the oxygen atom of a primary methoxy group of the host molecule.

4-chlorophenoxyacetic Acid (4-CPA)

4-CPA /β-CD Crystal Structure

The crystal structure determination of the 4-CPA/*β*-CD has given the following results (Tsorteki et al., 2005)**:** two guest molecules are encapsulated inside the dimeric cavity and thus the host:guest stoichiometry is 2:2. The guest molecules have been found disordered over two sites (A and B). The phenyl group of the guest forms angles of 50° and 60° (site A and B respectively) with the mean plane of the glucosidic O4*n* atoms. The two guest molecules are encapsulated inside the *β*-CD dimeric cavity with the planes of their phenyl groups forming a dihedral angle of 15.3° (site A) or 9.1° (site B) while the centroids of their phenyls are shifted by 4.17 Å (site A) and 3.62Å (site B). Therefore, the guest molecules are related by π-π interactions. One water molecule is also encapsulated inside the cavity hydrogen bonded with a primary hydroxyl group. The carboxyl groups of the guests are lying at the primary hydroxyl group's rim and they are hydrogen bonded with water molecules located at the interdimeric free space and with primary hydroxyl groups of a vicinal dimer. Moreover, hydrogen bonds are formed between the primary hydroxyl groups of the host and of a vicinal complex. The Cl atoms are located at the secondary hydroxyl groups region (Figure 10).

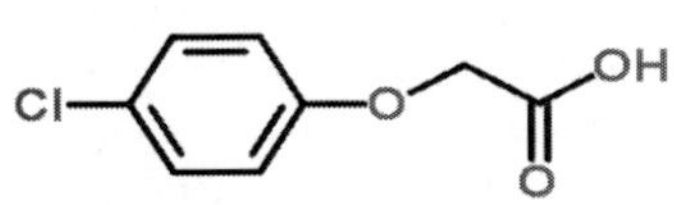

Figure 9. Scheme of the 4-chlorophenoxyacetic acid (4-CPA) molecule.

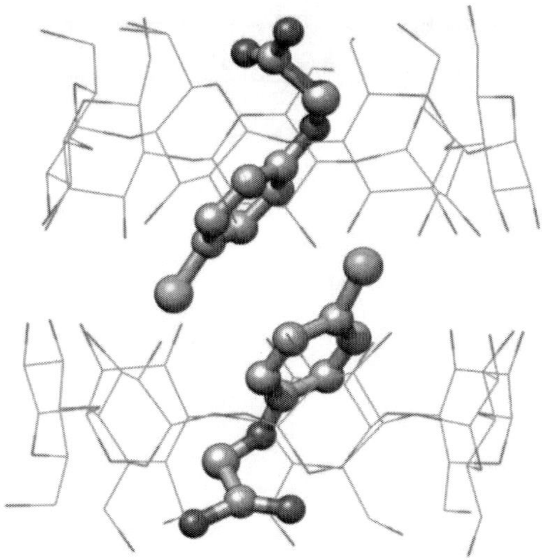

Figure 10. Crystal structure of the 4-CPA/β-CD complex. The formed β-CD dimer hosts two 4-CPA molecules disorder over two occupied site each. For clarity only one occupied site (A) of each and no water molecules are shown.

4-CPA/TRIMEB Crystal Structure

The 4-CPA/TRIMEB complex crystallizes also in the space group $P2_1$ but its crystal structure is quite unique (Tsorteki et al., 2005). The asymmetric unit contains two crystallographically independent complexes (ComplexA and ComplexB) and 3.37 water molecules distributed over 4 sites and hydrogen bonded with some atoms of the host and guest molecules. The existence of two symmetrically independent complexes within the asymmetric unit is due to the different orientation of the guest molecules inside the hosts' cavities. The host molecule of ComplexA is significantly distorted as one glucosidic residue is raised up like a hatch allowing the 'tail' of the guest molecule to protrude away from the cavity (Figure 11a). The phenyl ring of the guest forms an angle of about 40° with the mean plane of the O4n glucosidic atoms of the host. On the contrary, the phenyl ring of the guest molecule inside the host of ComplexB is almost perpendicullar to the O4n glucosidic atoms of the host and the guest molecule is 'swallowed' inside the cavity not protruding from it (Figure 11b). In ComplexB the guest is fully included axially in the hydrophobic cavity as in the case of the 4-CPA/β-CD complex whereas in ComplexA the guest is found laying 'equatorially' in the cavity although the conformation of the guest is the same with that of ComplexB. It is therefore noted that guests having the same conformation do not inevitably cause a definite conformation of the host. The crystal packing consists by complexes stacking along the *a*-axis forming antiparallel and crystallographically independent columns.

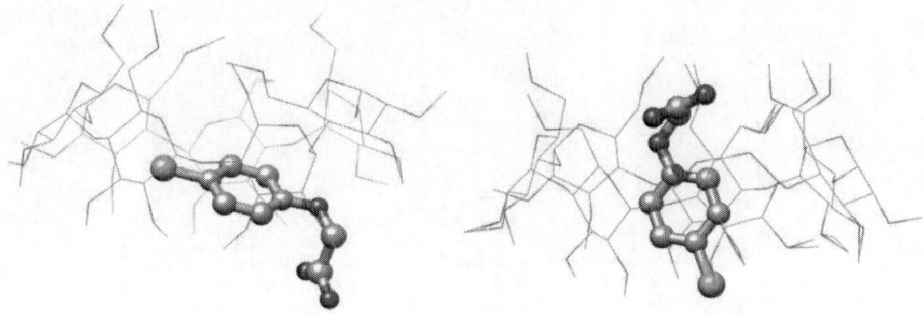

Figure 11. Crystal structure of 4-CPA/TRIMEB complex. Two different conformations of the complex are observed within the asymmetric unit (**a**) ComplexA (**b**) ComplexB.

2-methyl-4-chlorophenoxyacetic Acid (MCPA)

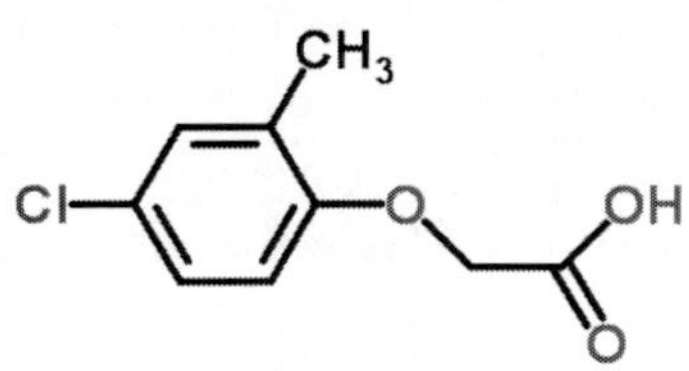

Figure 12. Scheme of the 2-methyl-4-chlorophenoxyacetic acid (MCPA) molecule.

MCPA/β-CD Crystal Structure

The MCPA/*β*-CD complex crystallizes in the space group $P2_1$. Its asymmetric unit contains two *β*-CD molecules (HostA and HostB) forming a head to head dimer via the usual hydrogen bonds between their secondary hydroxyl groups. The *β*-CD dimers are stacked along the crystallographic *a*-axis according to the Channel packing mode (Figure 13) (Mentzafos et al., 1991). Inside the *β*-CD dimer two disordered MCPA guest molecules occupying overall five sites are hosted (host:guest stoichiometry 2:2; Figure 13). The guest molecules are found highly disordered due to the nature of the weak intermolecular interactions between the host and the guest molecules (hydrophobic, Van der Waals, C–H...π and C–H...O interactions) as usually observed in the CD complexes. Moreover, in this case the guest molecule

(MCPA) has a torsional degree of freedom (C1–O–C8–C9: torsion angle) which allows to adopt a number of different conformations as found in the five occupation sites (Figure 15). In all occupied sites the guest molecule is accommodated with the same orientation its Cl atom being located at the interdimeric interface and its carboxyl group at the primary hydroxyl rim. The phenyl group of the guest forms an angle of about 60° with the mean plane of the glucosidic O4*n* atoms. Furthermore, the encapsulated pair of MCPA molecules inside the dimeric cavity is found π-π stacked.

In addition, 16.13 water molecules distributed over 36 sites are found in the asymmetric unit. The water molecules, except one that is found entrapped inside the dimeric cavity, form the usual hydrogen bonds network, contributing significantly to the stability of the crystal structure. It is rather rare to observe entrapped water molecules within the hydrophobic CD dimeric cavity (Giordano et al., 2001).

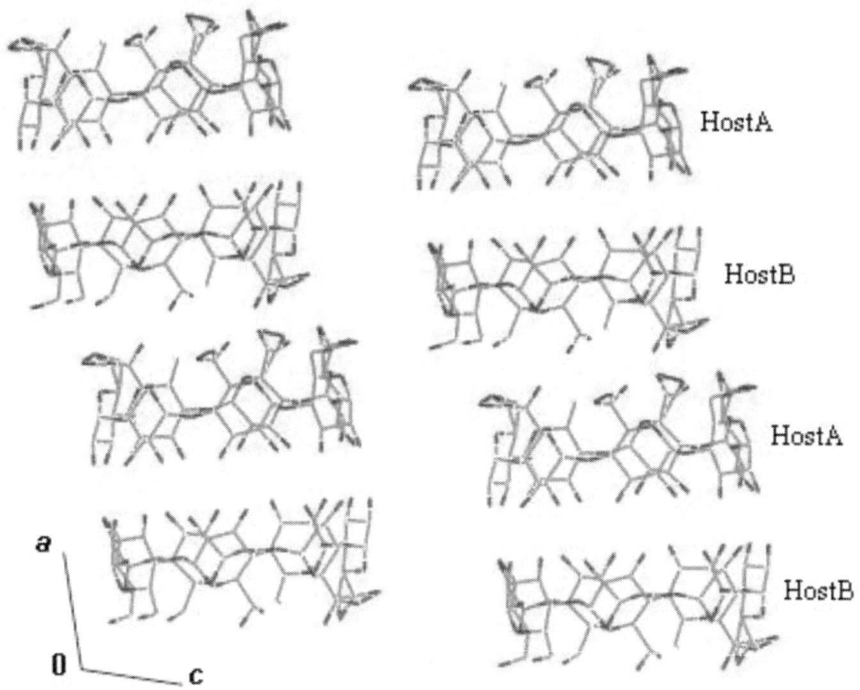

Figure 13. The channel developed along the *a*-axis. The shifting of the successive dimers of the same channel is 2.27 Å.

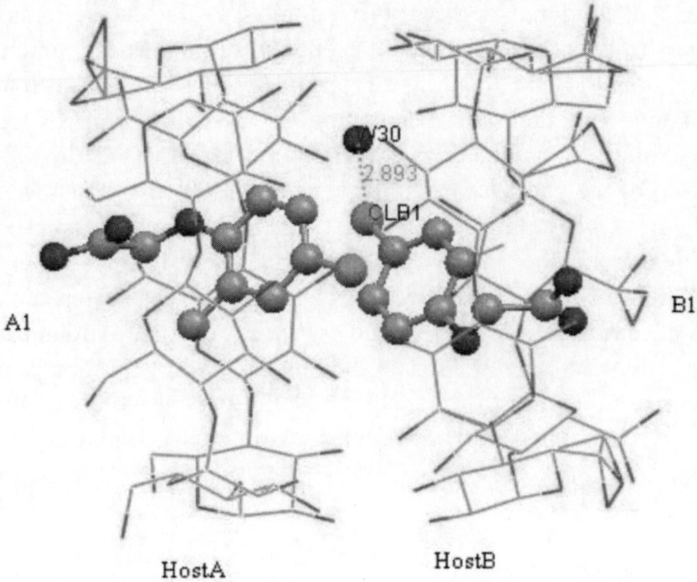

Figure 14. Crystal structure of the MCPA/β-CD complex. For clarity only the occupied sites A1 and B1 of the guest molecules and no water molecules are shown.

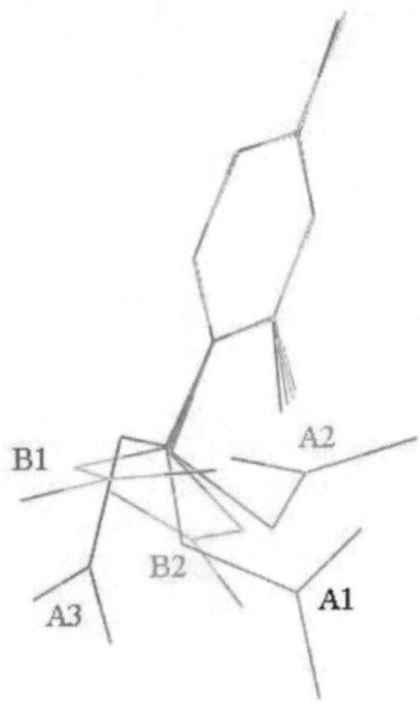

Figure 15. Superposition (alignment on the phenyl- ring) of the five different conformations adopted by MCPA in the five occupation sites of encapsulation in the MCPA/β-CD complex.

In all the cases where a water molecule has been found inside the dimeric cavity of the *β*-CD dimers, this water molecule is hydrogen bonded with either nitrogen or oxygen atoms of the guest molecule. In the case of the MCPA/*β*-CD complex the Chlorine atom of the MCPA occupying a particular site entraps a water molecule within the dimeric cavity via a Cl...H–O bond (Triantafyllopoulou et al., 2012).

MCPA/β-CD Infrared Spectroscopy (FTIR)

The infrared spectra of MCPA, *β*-CD, their physical mixture and their inclusion complexes obtained with CE and KN methods have been given by Garrido et al. (Garrido et al., 2012). The FTIR spectrum of MCPA shows a characteristic absorption band at 1744 cm^{-1} assigned to carboxylic C=O stretching vibration; three bands at 1597, 1495 and 1450 cm^{-1}, due to C=C of the aromatic ring; a band at 1427 cm^{-1} assigned to CH_2 (OCH_2 group) scissoring vibration; a band at 1298 cm^{-1}, corresponding to the C-O single bond stretching vibration; a wide band due to hydroxyl O-H of the carboxylic acid group appears at 1245 cm^{-1}.

In the physical mixture (PM), no variations on these wave number values were observed. Nevertheless, the spectra of the inclusion complexes obtained by CE and KN methods showed the displacement of the band corresponding to C=O stretching vibration, 1738 and 1728 cm^{-1} respectively, accompanied by a diminution of the band intensity. Moreover, the band at 1427 cm^{-1}, assigned to CH_2 (OCH_2 group) scissoring vibrations present a significant diminution on its intensity and a shift towards lower wave numbers. The overall FTIR data highlighted the formation of an inclusion complex in which MCPA is probably located inside of the ring of *β*-CD, because significant intermolecular hydrogen bonding interactions were detected between the acetic acid side chain of the herbicide and the CD. In fact, the FTIR spectra reveal important changes in some vibrational modes as consequence of the host/guest interactions.

MCPA/β-CD, Differential Scanning Calorimetry (DSC)

The physicochemical characterization of the MCPA/ *β*-CD complex using the DSC technique has been presented by Garrido et al. (Garrido et al., 2012). From the DSC profiles of pure components (MCPA, *β*-CD) and binary systems it is shown that:

i) The DSC curve of *β*-CD shows a broad endothermic signal around 78 °C that could be associated to the loss by evaporation of the water molecules existing as residual humidity. The broad endothermic effect occurring at

311.9°C is related to the melting of β-CD. ii) The thermal curve of MCPA is typical of a crystalline anhydrous substance with a sharp fusion endotherm at 119.8 °C, corresponding to the melting point of the compound, followed by an exothermal effect at 242.7 °C, attributed to its thermal decomposition. (iii) In the physical mixture (PM) DSC curves, it was possible to clearly distinguish the MCPA fusion endothermic peak around 119.8 °C in addition to the β-CD dehydration process. The obtained thermograms were in fact a superimposition of the individual components showing the absence of an interaction or a low yield of complexation between the herbicide and β-CD. (iv) The thermal curve obtained for the solid inclusion complex of MCPA with β-CD obtained by co-evaporation method does not show the MCPA endothermic peak at 119.8°C. Its absence could be explained in view of the formation of a complex between MCPA and β-CD. Considering KN systems, the absence of the MCPA melting endothermic peak was only noticed for MCPA/β-CD stoichiometric ratio of 1:2. Therefore, it was assumed that this method yield a 1:2 MCPA/β-CD complex, because this stoichiometry would allow maximum contact between the hydrophobic portion of the organic substrate with the apolar cavity of β-CD.

MCPA/β-CD, HP-β-CD and Me-β-CD, Phase Solubility Studies

Phase solubility studies by Garrido et al. (Garrido et al., 2014) examined the increase of the MCPA solubility in the presence of increasing amounts of β-CD, HP-β-CD and Me-β-CD. In all cases, A_L-type solubility diagrams were obtained as the MCPA solubility increased linearly with the increment of β-CD, HP-β-CD and Me-β-CD concentration, according to the classification established by Higuchi and Connors (Higuchi & Connors, 1965). The diagrams obtained indicate that the solubility of the MCPA is significantly increased by the presence of the cyclodextrins. Since the slope of the diagrams was less than 1, the complex stoichiometry was assumed to be 1:1. The stability constants, Ks, calculated for β-CD, HP-β-CD and Me-β-CD were 162 ± 12, 334 ± 9 and 398 ± 13 M^{-1}, respectively. The fact that the estimated stability constants for HP-β-CD and Me-β-CD systems are double of that for the β-CD indicates that these complexes are more stable. Garrido et al. attributed these complexes stability differences to the low polarity of the 2-hydroxypropyl or methyl substituents that can be more favourable to the formation of the complex. However, it should not be forgotten the increased flexibility and thus tunability of the HP-β-CD and Me-β-CD cavities that plays an essential role in the host's encapsulation ability as it has been shown by the crystallographic studies of similar complexes.

MCPA/β-CD, HP-β-CD and Me-β-CD, NMR Analysis

^{1}H NMR data was acquired by Garrido et al. (Garrido et al., 2012; 2014) to confirm the formation of the inclusion complexes of the MCPA with *β*-CD, HP-*β*-CD and Me-*β*-CD. The results showed that, in general, significant chemical shift variations are observed. The proton signals are upfield shifted with respect to the native compounds indicating host-guest interactions.

A detailed examination of the spectra of the MCPA/*β*-CD complexes indicates that the experimental shifts of the *β*-CD cavity protons have a noticeable displacement. The observed modification of the protons present in the *β*-CD inner cavity can be justified by the complex formation with inclusion of MCPA and thus to the presence of ring current effects generated by the aromatic nucleus of MCPA that in the inclusion process affect the neighboring proton signals of the host. The proposed guest position in the cyclodextrin given by this NMR analysis is one with the Chlorine atom of the MCPA locating at the rim of the primary hydroxyls of the *β*-CD. The proposal is consistent with the upfield shifts of the *β*-CD cavity protons as they experienced anisotropic shielding due to the inclusion of the MCPA aromatic ring. However, this orientation of the guest molecule inside the hydrophobic *β*-CD cavity is opposite to the one found in the crystal structure of the complex. It is possible that the crystal contacts are responsible for the formation of the head to head *β*-CD dimer which allows the specific encapsulation of the MCPA molecules and the π-π interactions between them in the crystalline state.

The stoichiometry of complexes was also determined by comparing the ^{1}H NMR integration data. For the MCPA/*β*-CD complex obtained using the kneading (KN) method, a stoichiometry of 1:2 was found while for MCPA/*β*-CD complex obtained using the co-evaporation (CE) method the stoichiometry was 1:1. These results can be interpreted on the basis of the different techniques used to prepare the CD complexes. The experimental results obtained for a CD complex can depend on the technique used since the CD complexes are held together by weak forces. Moreover, the complexation process depends heavily on the experimental conditions (pH, cosolvent, temperature, etc.). Thus, the complexes can involve species of different stoichiometries, depending on the technique used, in this case CE and KN methods (Garrido et al., 2012).

Similar results were obtained for the complexes MPCA/HP–*β*–CD and MCPA/Me–*β*–CD (Garrido et al., 2014). Significant upfield chemical shift variations of proton signals in the complexes with respect to the native compounds have been detected. Noticeable upfield shift displacements were also observed for OCH_2 group and the H6 aromatic proton of MCPA. The stoichiometry of the complexes has been also determined by comparing the ^1H NMR integration data between the H6 proton in MPCA and H1′ proton of the corresponding CD. For the MPCA/HP–*β*–CD and MPCA/Me–*β*–CD complexes a stoichiometry of 1:1 was obtained (Garrido et al., 2014).

MCPA/β-CD; MCPA/7Me-β-CD; MCPA/14Me-β-CD, in Silico Studies

Molecular mechanics simulations have been performed by Garrido et al. (Garrido et al., 2014) to understand the increased solubility/binding of MCPA with modified (methylated) *β*-CDs. For this study, in addition to the native *β*-CD, the following two different methylated *β*-CDs were used: 7Me-*β*-CD, with one methyl group per anhydroglucose unit and 14Me-*β*-CD (heptakis (2, 6-di-O-methyl)-b-cyclodextrin or DIMEB), with two methylgroups per anhydroglucose unit.

In order to calculate the binding energy differences, the free energy of solvation of seven different systems were computed, namely: MCPA, *β*-CD, 7Me-*β*-CD, 14Me-*β*-CD, MCPA/*β*- CD, MCPA/7Me-*β*-CD and MCPA/14Me-*β*-CD. Initially, systems were solvated inside a cubic box of water with ~4 nm side and then 7 nanoseconds of NPT molecular dynamics simulations were run with 1 femtosecond time steps at room temperature. Finally, the energetics of snapshot conformations were analyzed at 1 picosecond intervals with an implicit solvent model. The lowest energy configurations were obtained for MCPA/*β*-CD and MCPA/14Me-*β*-CD complexes. MCPA/7Me*β*-CD and MCPA/14Me-*β*-CD complexes yielded a 1.8 kcal/mol increase in binding energy with respect to MCPA/*β*-CD. Moreover, differences in the orientation of the encapsulated MCPA were observed: for MCPA/*β*-CD the carboxylic group sticks out from the narrower side of *β*-CD while for MCPA/14Me-*β*-CD the carboxylic group points to the opposite direction. According to these results, MCPA flips orientation when bound to modified *β*-CDs increasing the stability of the resulting complexes. The prediction of the MCPA orientation in the MCPA/*β*-CD complex is in accordance with the one found in the determined crystal structure of the complex (Triantafyllopoulou et al., 2012) but it is opposite to the one assumed by the findings of the ^{1}H NMR studies of the complex (see previous paragraph: "MCPA/β-CD, HP-β-CD and Me-β-CD, NMR analysis").

2,4-Dichlorophenoxyacetic Acid (2,4-D)

2,4-D/α-CD, Phase Solubility Studies

The formation of the 2,4-D/α-CD inclusion compound was studied in the solution state via the phase-solubility diagram by Pérez-Martinez et al. (Pérez-Martinez et al., 1998). The samples were analyzed spectrophotometrically at 284 nm. The obtained diagram was classified as of Bs type according to Higuchi and Connors (Higuchi & Connors, 1965). An insoluble microcrystalline complex was formed from the solution at higher α-CD concentrations. The diagram shows a plateau region before the descending part of the curve. From the length of the plateau it is possible to estimate a 1:2 stoichiometry of the 2,4-D/α-CD complex. Thus, it is assumed that according to the findings in the plateau region, one molecule of 2,4-D is entrapped by two molecules of α-CD because of the small cavity size of the latter. However, the increase of the solubility estimated from the initial inclined line portion is significantly different from that at high α-CD concentrations. This result was interpreted in terms of the initial formation of a 1:1 complex. Therefore, Pérez-Martinez et al. assumed the formation of two or more distinct complexes in the system, one or two of these being responsible for the initial rise in solubility, while another is precipitated later.

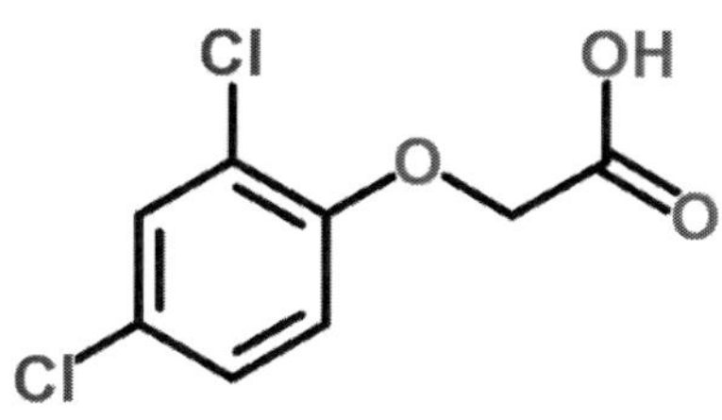

Figure 16. Scheme of the 2,4-Dichlorophenoxyacetic acid (2,4-D) molecule.

2,4-D/α-CD, Differential Scanning Calorimetry (DSC)

Although the results obtained from the solubility study strongly indicate the formation of a true inclusion compound of 2,4-D with α-CD, Pérez-Martinez et al. (Pérez-Martinez et al., 1998) analyzed the final solid product with the DSC technique in order to investigate if it was inclusion complex or just a physical mixture. DSC records of the pure components, the physical mixture and the kneaded, co-precipitated and spray-dried systems were presented. By analyzing these curves, it is revealed that 2,4-D forms an inclusion compound with α-CD only in the cases of co-precipitate and spray-dried samples. These complexes present a 2,4-D:α-CD stoichiometry of 1:2.

2,4-D/β-CD Crystal Structure

X-ray crystallographic studies (Triantafylopoulou, 2013) have shown that the complex 2,4-D/β-CD crystallizes in the space group *P*1 forming head to head dimers (Figure 17). The dimers consist of two host molecules (HostA and HostB) stacking according to the channel (*CH*) packing mode. 1.7 water molecules, distributed over five sites, have been found entrapped inside the dimeric cavity, together with one guest molecule distributed over three sites. Therefore, the host:guest stoichiometry is 2:1 and three models are possible.

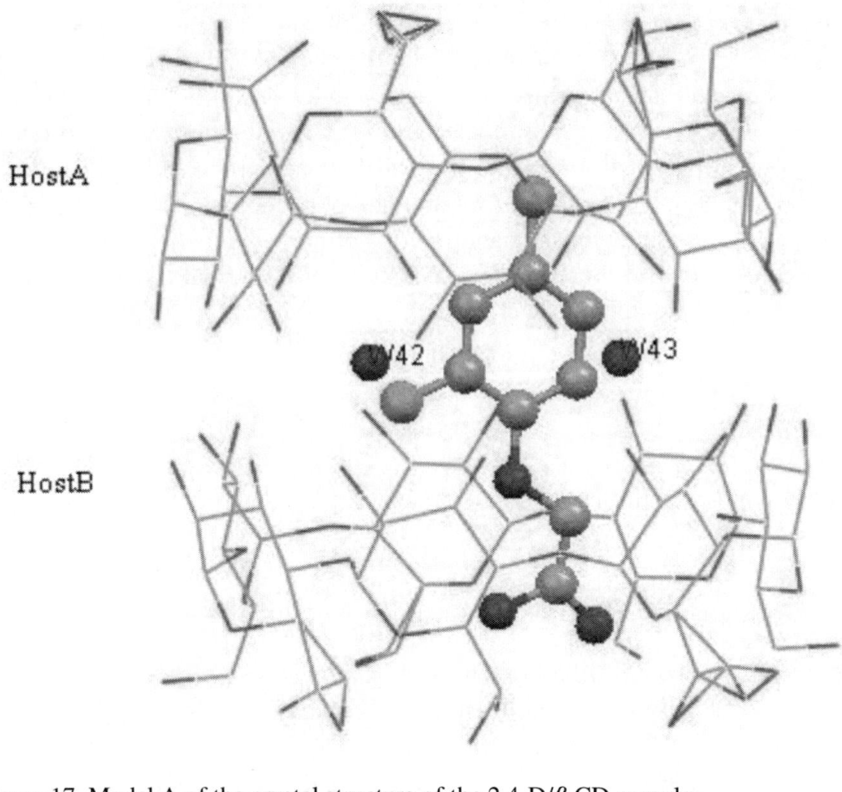

Figure 17. Model A of the crystal structure of the 2,4-D/β-CD complex.

Model A: The guest molecule is accommodated inside the cavity in such a way that it is left enough space for two water molecules to be encapsulated inside the dimer. These water molecules are accommodated at the dimeric interface where the Cl atom of the guest is also located and they are found hydrogen bonded between them and also with the chlorine atom. The other Cl atom of the guest lies at the mean plane of the O4*n*B atoms. The carboxyl group of the guest is located inside the primary hydroxyl groups region and its phenyl group is nearly perpendicular to the O4*n*A and O4*n*B mean planes of the HostA and HostB.

Model B: Once again, two water molecules are found entrapped inside the dimeric cavity. However, the orientation of the guest molecule inside the cavity is the opposite to that of the Model A and there is enough space for the two water molecules to be included inside the cavity without being hydrogen bonded with any atom of the guest.

Model C: The main part of the guest molecule is accommodated inside the HostA cavity leaving the HostB cavity practically empty. This allows three water molecules to be entrapped inside the dimeric cavity, two of them being hydrogen bonded with the chlorine atoms of the guest molecule (Figure 18a). The phenyl group is located at the dimeric interface whereas the carboxyl group located at the primary hydroxyl groups' region is hydrogen bonded with the primary hydroxyl group of the vicinal dimer of the channel (Figure 18b).

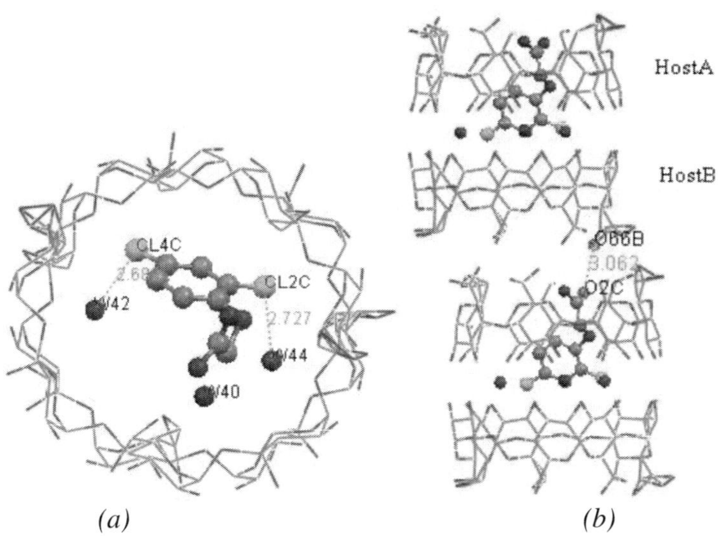

Figure 18. (a) Model C of the crystal structure of the 2,4-D/β-CD complex (top view). (b) Crystal packing of the 2,4-D/β-CD complex.

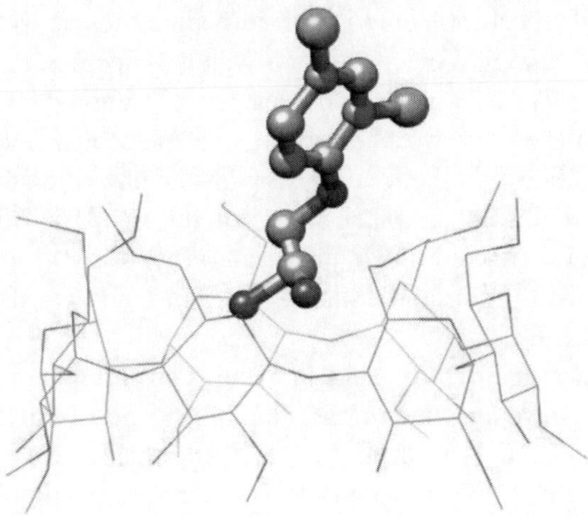

Figure 19. Crystal structure of the 2,4-D/DIMEB complex. Water molecules are omitted for clarity.

2,4-D/DIMEB Crystal Structure

Only two auxin/DIMEB complexes have been elucidated by X-ray crystal structure analysis. They both crystallize in the space group $P2_12_12_1$ according to a head-to-tail mode and a unique molecular packing.

The 2,4-D/DIMEB crystal structure has been determined by Tsorteki and Mentzafos (Tsorteki and Mentzafos, 2002). The complexes were found stack along the *a*-axis and their molecular packing is characterized as a puckered screw channel (*SC*) (Mentzafos et al., 1991). The dichlorophenoxy group of the guest is located outside the hydrophobic cavity at the space existing among the host molecules (Figure 19). Only the guest's carboxyl group penetrates into the host cavity from the primary methoxy groups rim as the gust's dichlorophenoxy group cannot fit inside the rigid cavity. The primary methoxy groups block the cavity of the vicinal host of the screw channel from its secondary rim. Inside the cavity a water molecule is entrapped, hydrogen bonded with the carboxyl group of the guest.

2,4-D/TRIMEB Crystal Structure

The 2,4-D/TRIMEB crystal structure has been determined by Tsorteki et al. (Tsorteki et al., 2004). The crystal packing of the complex has many common features with that of the IBA/TRIMEB complex as both complexes crystallize in the space group $P2_1$ stacking along the *a*-axis. In the case of the 2,4-D/TRIMEB two water molecules are found encapsulated inside the host cavity. The guest molecule is found disordered over two sites. The mean plane of the phenyl ring of both sites is nearly perpendicular to the mean plane of the glucosidic O4*n* atoms (Figure 20). The carboxyl group of the disordered guest protrudes from the secondary methoxy groups region hydrogen bonded with water molecules. One water molecule is hydrogen bonded with the oxygen atom of a secondary methoxy group of an adjacent host molecule participating in the crystal contacts.

2,4,5-Trichlorophenoxyaceticacid (2,4,5-T)

2,4,5-T/β-CD and 2,4,5-T/γ-CD. Formation and studies of the complexes in aqueous solutions.

The formation of 2,4,5-T complexes with β- and γ-CD (2,4,5-T/β-CD; 2,4,5-T/γ-CD) and the effect of this process on the rate of the steady-state photolysis of the 2,4,5-T has been studied by Yurkova et al. (Yurkova et al., 2011; 2012). The experiments were carried out in weakly basic aqueous solutions (pH 9-10) within the range of acid concentration amounting to 10^{-5}–10^{-3}mol/L.

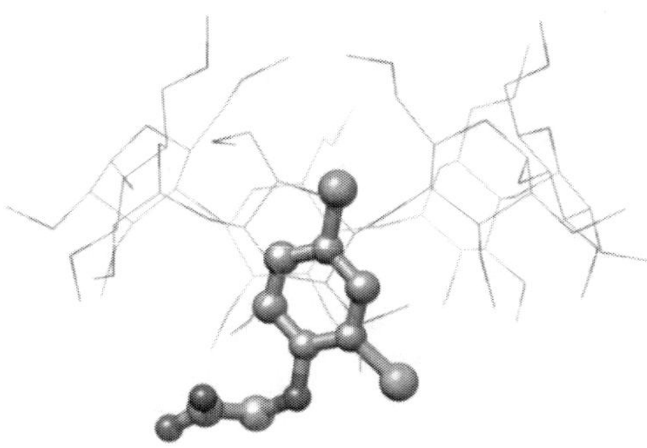

Figure 20. Crystal structure of the 2,4-D/TRIMEB complex. For clarity, only one occupied site and no water molecule is shown.

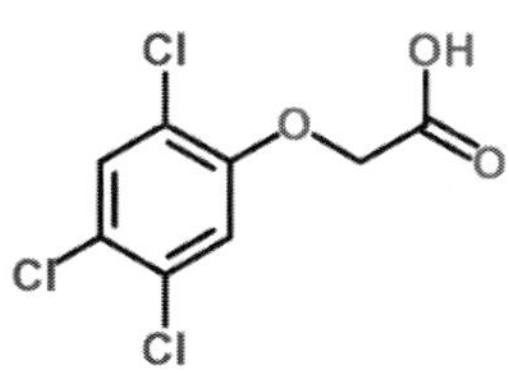

Figure 21. Scheme of the 2,4,5-Trichlorophenoxyaceticacid (2,4,5-T) molecule.

The concentration of 2,4,5-T was determined at λ=289 nm. Adding the cyclodextrins resulted a slight change of the shape of the 2,4,5-T spectrum, which could indicate a complexation reaction to occur. In order to determine the stability constant of the inclusion complex, Yurkova et al. registered the absorption spectra of 2,4,5-T aqueous solutions under varying the concentration of CD. The values obtained was 1100 and 150 mol/L for 2,4,5-T/β-CD and 2,4,5-T/γ-CD respectively. Finally, by investigating the steady-state photolysis of 2,4,5-T and its complexes with CD in aqueous solutions they concluded that in the presence of β-CD and γ-CD, there is a changing in the rate of 2,4,5-T photodegradation.

Formation of the Ternary 2,4,5-T/Fe (II)/β-CD Inclusion Complex in the Gas Phase

Electrospray mass spectrometry (ES-MS) is an important tool for characterization of noncovalent binding in the gas phase. In the study of Cai et al. (Cai et al., 2003), iron (II) has been introduced as a dication to enhance the detection of *β*-CD plus 2,4,5-T complex in ES-MS. The results of the ES-MS and ES-MS/MS experiments suggested that it is the polar portion of 2,4,5-T (i.e., the carboxylic acid moiety) that is favored for inclusion in the *β*-CD cavity, rather than the non-polar aromatic part. The polar part of such molecules can interact with Fe^{2+} inside the CD cavity via (oxygen electron) electrostatic interactions, even when the nonpolar part of the aromatic compound is situated outside the CD cavity. This means that complex formation between cyclodextrin, iron (II) and guest molecule in the gas phase may be different from that which is favored in aqueous solution where generally the non-polar part of a molecule would be driven into the CD cavity by hydrophobic forces. Moreover, the electrostatic interactions between the polar moiety of the guest molecule and the hydroxyl groups of cyclodextrin might stabilize a guest located outside the CD cavity.

2,4,5-T/TRIMEB Crystal Structure

All the attempts that were made to crystallize the inclusion complexes of 2,4,5-T with native *β*-CD or DIMEB failed. Only the 2,4,5-T/TRIMEB complex gave suitable crystals for X-ray diffraction and the crystal structure of that complex was successfully determined (Triantafyllopoulou et al., unpublished data). The complex was found crystallizing in the space group $P2_12_12_1$. The asymmetric unit contains one host molecule; one guest molecule and 1.8 water molecules distributed over four sites. In the crystalline state the complexes are accommodated along the twofold screw *b*-axis forming a Screw

Channel. The phenyl ring is accommodated inside the hydrophobic cavity of the TRIMEB oriented in such a way that its two Chlorosubstituents and the carboxyl group protrude from the cavity (Figure 22). Both the oxygen atoms of the carboxyl group are hydrogen bonded with water molecules located at the interface of the Channel.

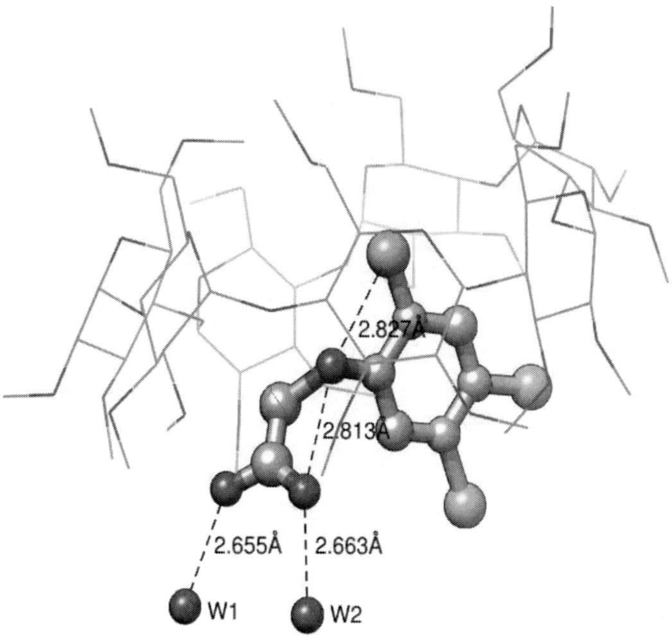

Figure 22. The crystal structure of the 2,4,5-T/TRIMEB complex. For clarity the water molecules are omitted.

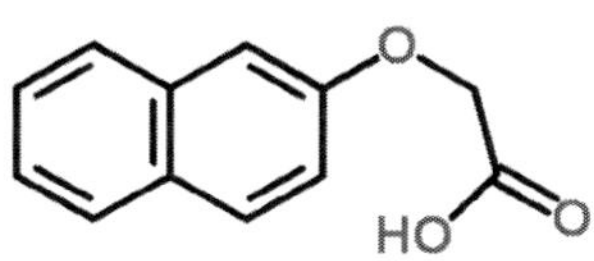

Figure 23. Scheme of the 2-napthoxyaceticacid (2-NOA) molecule.

2-napthoxyaceticacid (2-NOA)

2-NOA/β-CD Crystal Structure and NMR Studies

X-Ray crystallography along with ^{1}H NMR studies of the 2-NOA/β-CD inclusion complex have been presented by Kokkinou et al. (Kokkinou et al.,

2001). The determination of the crystal structure of the complex revealed a dimeric complex with host:guest stoichiometry of 2:1 meaning that only one guest molecule is included inside the cavity of each dimer though there is enough space for two guest molecules to be encapsulated inside the cavity (Figure 24). The ^{1}H NMR spectra in an aqueous solution indicated a mixture of two complexes having 2:2 and 2:1 host:guest stoichiometry respectively justifying the large void observed inside the dimeric cavity of the crystal structure of the complex. The guest molecule located inside the cavity forms an angle of 63° with the glycosidic O4*n* mean plane. Its carboxyl group is located at the primary side of the host and its oxygen atoms are hydrogen bonded with primary hydroxyl groups either of their own host or of the vicinal *β*-CD.

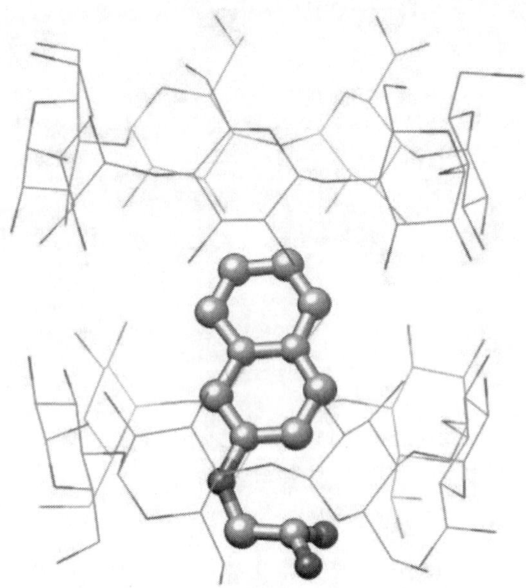

Figure 24. The crystal structure of 2-NOA/*β*-CD complex. Water molecules are omitted for clarity.

2-NOA/β-CD in Aqueous Solution

The inclusion complex of 2-NOA/*β*-CD has been investigated in aqueous solution by absorptiometric and spectrofluorimetric studies (Munoz de la Pena et al., 1993). By the Scatchard plot of the measured fluorescence intensities (Connors et al., 1987) a 1:1 guest:host stoichiometry of the 2-NOA/*β*–CD complex was determined and an association constant of 560 ± 110 M^{-1} was calculated. By examining the steric constraints in the formed complex it was proposed a mode of inclusion according to which the naphthalene of the 2-NOA is accommodated axially inside the *β*–CD cavity. This is verified by the X-ray crystallographic studies (Kokkinou et al., 2001). However, the orientation of the encapsulated 2-NOA inside the *β*–CD dimer as determined by the crystal structure of the complex is the opposite of that proposed by Munoz de la Pena et al. Moreover, in the crystalline state, 2-NOA was found immersed deep inside the dimeric cavity its carboxyl group not protruding from the primary hydroxyls rim of the *β*–CD (Figure 24).

2-NOA/β-CD, in Silico Studies

2-NOA/*β*-CD has been used as a model system for docking studies (Ermondi et al., 2007). In those studies, instead of performing a generic conformational analysis, eight *β*-CD models were prepared by modulating the shape of the cavity which tends to become oval in the presence of ligands and the position of the primary oxygens. The docking results showed that for all models, the molecules of 2-NOA were found in the interior of the CD cavity located in two main poses: one, in agreement with the X-ray data with the naphthyl group in the vicinity of the CD secondary oxygens and the other in the opposite situation with the polar moiety in the vicinity of CD primary oxygens. These results are in line with NMR data (Kokkinou et al., 2001) indicating that, in solution, both these types of 1:1 complexes are found. Finally, since the docking procedure requires final minimization of the complex, the results obtained from the eight models were merged into one database and sorted by increasing the total energy of the complex. Analysis of this database reveals that the complex with the lowest energy is the one with its ligand in the same orientation as in the crystallographic structure. However, the energy results suggested that, in CD/ligand systems, the role of the energy values in the prediction of log *K* is not clear, partly because the contribution of entropy has been neglected.

2-naphthaleneacetic Acid (2-NAA)

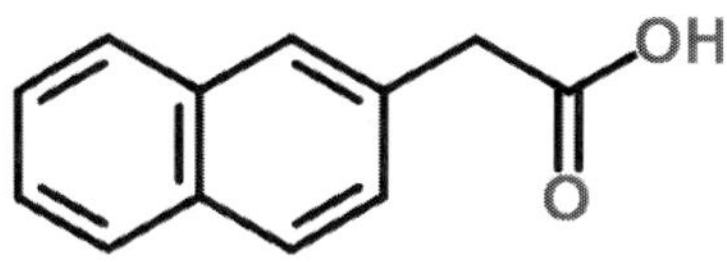

Figure 25. Scheme of the 2-naphthaleneacetic acid (2-NAA) molecule.

2-NAA/β-CD Crystal Structure

Once again *β*-CD dimers are formed in the crystalline state of the 2-NAA/*β*-CD inclusion compound. The complex has been found crystallizing in the *P*1 space group with two *β*-CD molecules, HostA and HostB, in the asymmetric unit forming a head to head dimer (Triantafyllopoulou et al., 2013). The dimers stack along the *c*-axis and their molecular packing is channel (*CH*). Inside the cavity of each host, a guest molecule is accommodated disordered over two sites, A1 and A2 inside the HostA, and B1 and B2 inside the HostB (Figure 26). Therefore, the host:guest stoichiometry is 2:2. The geometric analysis indicates that there can be two pairs of guest sites, the A1-B1 and the A2-B2, described by two possible models of the crystal structure. The distance between the centroids of the naphthalene planes of the first pair is 3.6 Å, the shifting 1.6 Å and the dihedral angle 1.3°, suggesting a π-π interaction. In the second pair the corresponding distances and angle are 3.6 Å, 2.0 Å and 12°, suggesting a slight overlap and a reduced π-π interaction. The naphthyl planes are inclined about 65° towards the mean plane of the glucosidic O4*n* atoms. The carboxyl groups protrude from the primary hydroxyl groups region. The carboxyl group of the site A1 is hydrogen bonded with that of the B1 site of the vicinal complex of the channel. The same hydrogen bond is observed between the A2 and B2 sites. As the guest molecules of the two pairs are connected by π-π interactions and they are also hydrogen bonded with the guest molecules of the adjacent complexes of the same channel we can say that in the crystalline state nanowires made of 2-NAA molecules are formed filling the interior hydrophobic space of the nanotubes made by the *β*-CD dimers (Figure 27).

2-NAA/TRIMEB Crystal Structure

The *2-NAA*/TRIMEB complex crystallizes in the space group $P2_12_12_1$ (Triantafyllopoulou et al., 2013). Its asymmetric unit consists mainly of one host molecule and one guest molecule distributed over two sites (Figure 28). The complexes stack along the *b* twofold screw axis forming a screw channel (*SC*). A water molecule is also found in the asymmetric unit located at the free space between the hosts of the *SC* and it is hydrogen bonded with an oxygen atom of the methoxy group of a vicinal complex of the *SC*. Both sites of the guest of the naphthalene moiety are accommodated in the secondary methoxy group's rim their carboxyl groups protruding from the cavity and being hydrogen bonded with the water molecule existing in the free space between the hosts of the channel. Thus, the sole water molecule found in the asymmetric unit bridges the host and guest molecules of the channel.

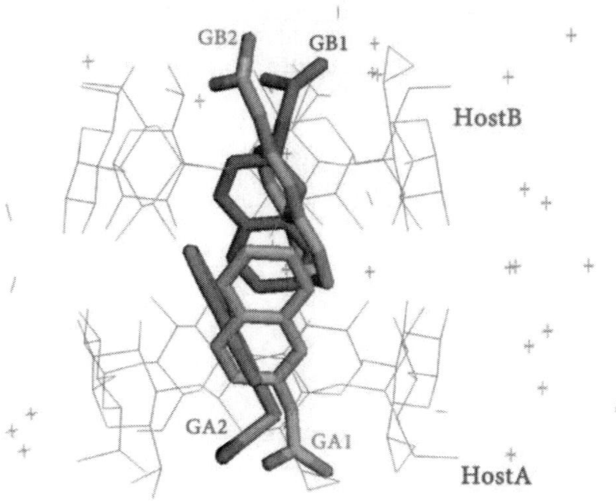

Figure 26. The asymmetric unit of the 2NAA/β-CD crystal structure. Two guest molecules are accommodated inside the formed head-to-head dimer occupying either the GA1–GB1 or the GA2–GB2 pair of sites.

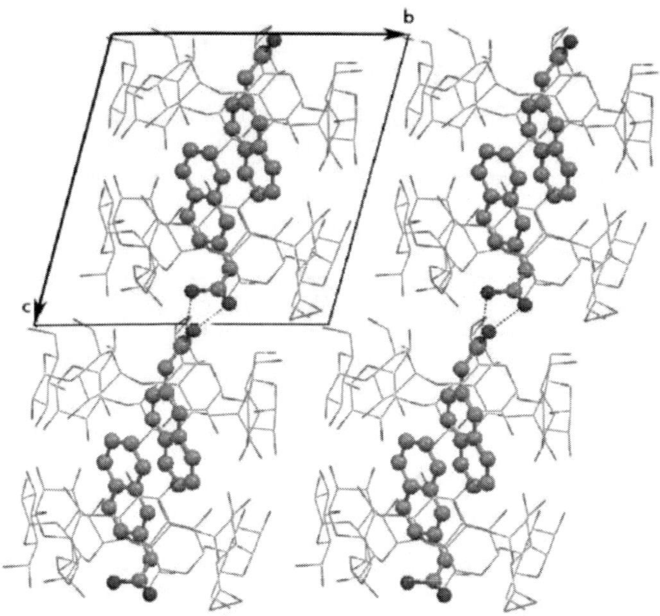

Figure 27. The type of the molecular packing of the 2NAA/β-CD crystal structure is that of a channel formed by dimers along the c axis. The guests found inside these nanotubes are linked by π–π interactions inside the dimeric cavities and by H-bonds between adjacent dimers forming nanowires.

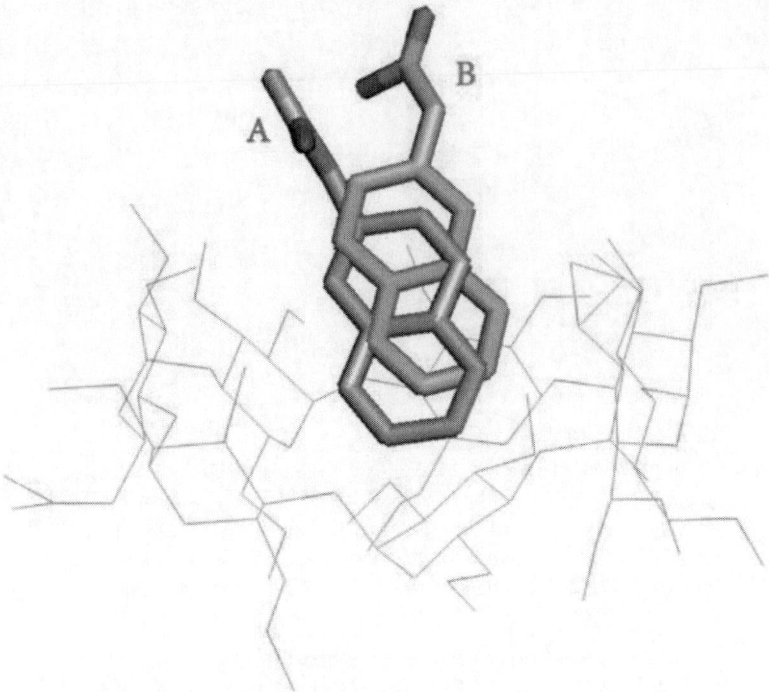

Figure 28. The 2NAA/TRIMEB crystal structure. The guest molecule is distributed over two sites, A and B, having occupancy factors 0.78 and 0.22 respectively.

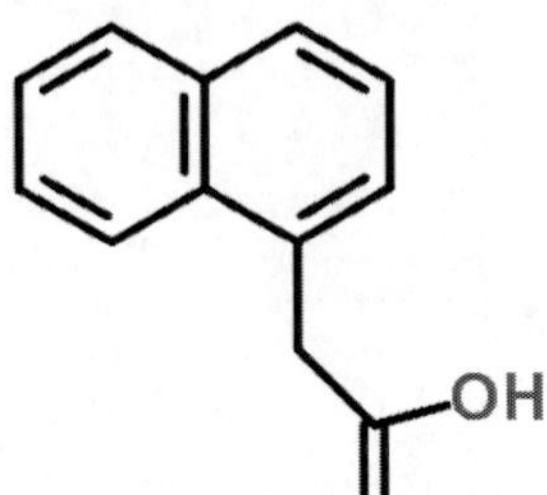

Figure 29. Scheme of the 1-naphthaleneacetic acid (1-NAA) molecule.

1-Naphthaleneacetic Acid (1-NAA)

1-NAA/β-CD Crystal Structure

Wang et al. have determined the crystal structure of a sandwich inclusion complex of β-CD with 1-NAA (Wang et al., 2011). This complex crystallizes in the space group *P*1. The host *β*-CDs form head to head dimers stacking along the *a*-axis their crystal packing being the *IM* molecular packing (Mentzafos et al., 2011). The crystallization of this complex was succeeded because of the high concentration of ethanol used in the crystal growth procedure. In particular, the solvent used was ethanol:water v/v = 1:2 and thus two ethanol molecules were encapsulated inside the hydrophobic *β*-CD cavities whereas the 1-NAA molecule was entrapped in a sandwich mode at the interface of the dimer (Figure 30). The 1-NAA molecule is accommodated at the dimeric interface its naphthyl plane being parallel to the O4*n* mean plane and its carboxyl group hydrogen bonded with a secondary hydroxyl group of the host. Each ethanol molecule is located inside a *β*-CD host monomeric cavity. The one encapsulated ethanol molecule has an horizontal orientation its hydroxyl group being hydrogen bonded with a water molecule whereas the other has a vertical orientation its hydroxyl group pointing to the secondary hydroxyl groups region. The difference between the *β*-CD inclusion complex with 1-NAA and that with 2-NAA as a guest (molecules that differ only in the position of the substitution in their naphthalene) is striking. 2-NAA is encapsulated axially inside the hydrophobic *β*-CD dimeric cavity whereas 1-NAA can only be entrapped in a sandwich mode at the *β*-CD dimeric interface when the cavities of the *β*-CD monomers are blocked by ethanol molecules. This difference in the inclusion ability of these two molecules inside the *β*-CD cavities may shed light on the mechanisms leading to the strong action of 1-NAA and the much weaker action of 2-NAA. It seems that the steric effects caused by the different position in substitution in the naphthalene ring play a crucial role on the binding ability and consequently the functionality of these auxins.

1-NAA/β-CD in Aqueous Solution

The inclusion complex of 1-NAA with *β*-CD has been investigated by absorptiometric and spectrofluorimetric studies in aqueous solution (Munoz De La Pena et al., 1993). It has been demonstrated that 1-NAA forms a 1:1 host:guest complex when included in the *β*-CD. In addition, an association constant of 100 ± 50 M^{-1} was calculated by the Scatchard plot of the measured fluorescence intensities (Connors et al., 1987). This value is lower than that calculated for the 2-NOA/*β*–CD complex indicating a weaker complexation caused by a possible partial inclusion of the 1-NAA into the *β*–CD cavity. It is therefore concluded that the steric effect of the position of the naphthalene substituents is so strong that the complexation with CDs is highly affected and the structures of the complexes should be different.

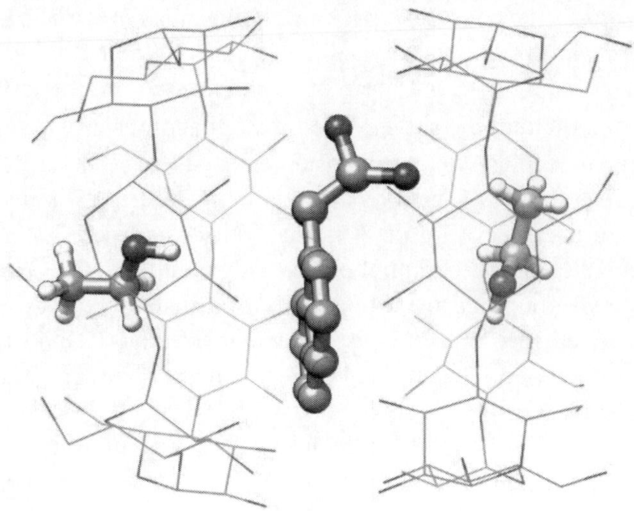

Figure 30. A sandwich-like inclusion mode of the 1-NAA in *β*-CD dimer.

1-NAA/DIMEB Crystal Structure

The crystal structure determination of the 1-NAA/DIMEB complex (Tsorteki et al., unpublished data) has revealed a unique crystal packing as the asymmetric unit consists of two hosts (HostA and HostB), three guests, having an overall occupation factor 2 and two water molecules. Therefore, the host:guest stoichiometry is 2:2. One guest molecule disordered over two sites is located outside the cavity having an overall occupation factor equal to 1. Two models are possible taking into account the distances between the guests and the water molecules and excluding the coexistence of some of the found occupied sites for steric reasons:

*Model A:*The cavity of the HostA contains one guest molecule disordered over two sites having an overall occupation factor 0.5 and the cavity of HostB includes a disordered water molecule.

Model B: Inside the cavity of the HostA a disordered water molecule is encapsulated and the HostB cavity includes a disordered guest molecule having an overall occupation factor 0.5.

It is noted that the guest molecules inside the cavities of the HostA and HostB have opposite orientations. The carboxyl group of the guest included in the HostA cavity points towards the rim of the secondary methoxy groups whereas the carboxyl of the guest included in the HostB points towards the primary methoxy groups region (Figure 31).

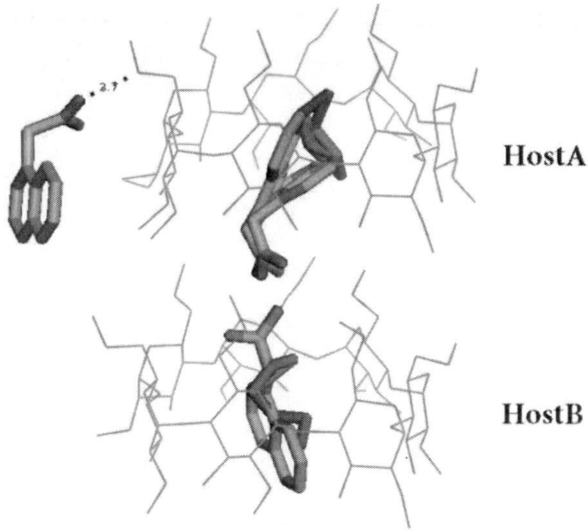

Figure 31. Crystal structure of 1-NAA/DINA.

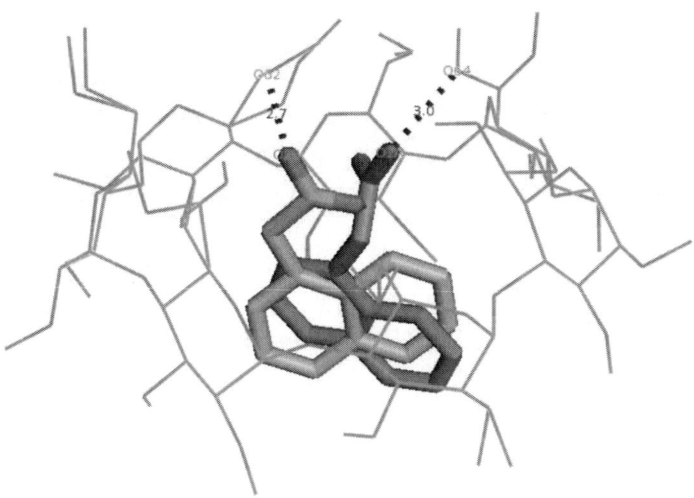

Figure 32. The crystal structure of the *1-NAA*/TRIMEB inclusion complex.

1-NAA/TRIMEB Crystal Structure

The 1-NAA/TRIMEB complex crystallizes in the space group $P2_1$ stacking along the shortest *a*-axis (Tsorteki et al., unpublished data). The crystal packing consists of antiparallel columns related by the twofold screw *b*-axis. The asymmetric unit contains one host and one guest molecule but no water molecules. The guest is found disordered over two sites. The naphthyl moieties of both sites are accommodated at the secondary methoxy groups rim being nearly perpendicular to the mean plane of the O4*n* glucosidic atoms. Their carboxyl groups are lying deeply inside the primary methoxy groups rim hydrogen bonded with the oxygen atoms of these methoxy groups (Figure 32). Compared to the other crystal structures of inclusion complexes of various auxins in TRIMEB, 1-NAA has the deepest immersion inside the TRIMEB hydrophobic cavity.

Other Applications

1-NAA Inclusion in β-CD Nanosponges

Recently new compounds, the nanosponges, have been obtained by hyper-reticulation of *β*-CD and seem to form more stable complex with growth regulators releasing them gradually. The effectiveness of 1-NAA inclusion in nanosponges as compared to *β*–CD has been assessed by testing the improvement of rhizogenesis in globe artichoke spring cultivar 'Romanesco C3' (Cavallaro et al., 2013). Significantly higher percentages of rooted plantlets were observed in the treatments where 1-NAA was supplied with *β*-CD (83%) or included in nanosponges (78%) as compared to the 65% registered in the medium where 1-NAA was used alone. It was thus concluded that the use of the water soluble complex 1-NAA/nanosponges may simplify the preparation of the rooting medium by eliminating the preliminary preparation of 1-NAA stock solutions and the separate addition of *β*-CD.

1-NAA in β-CD Hydrogels for pH Sensitive Release

Yang et al. prepared a *β*-CD hydrogel in which 1-NAA was loaded by taking advantage of its hydrophobic interaction with the cavity of *β*-CD in order to endow a pH sensitivity to the hydrogel (Yang et al., 2010). The pH dependency of the new hydrogel was tested by examining the release of blue dextran (a water-soluble dye) from it. The results showed that the hydrogels were pH sensitive in terms of release and swelling, when 1-NAA molecules were included in the cavities of cross-linked *β*-CDs. 1-NAA included in the cavities endowed the pH-sensitive properties to the *β*-CD hydrogels. The degree of release and the pH sensitivity of the hydrogels could be controlled by not only the amount of cross-linker used for the preparation of hydrogels but also the amount of 1-NAA loaded in the hydrogels. The hydrogels proposed in that study could be used as a delivery system, which releases its content in a pH-dependent manner.

1-NAA with β-CD Grafted Polyethyleneimine (βCD-PEI) Forming Microgels

Microgels were successfully prepared by taking advantage of the electrostatic interaction and the hydrophobic interaction of 1-NAA with *β*CD-PEI (Lee et al., 2011).

These microgels can be used as carriers for the pH-dependent release of water-soluble drugs. The results of the Lee et al. study showed that the shape of the microgel depends on the pH of the medium. In particular, under a strong acidic condition (e.g., pH 3.0) fibrous microgels were observed whereas globular microgels were found in the range of pH 4.0 to pH 8.0 and no microgels were observed at pH 9.0 and pH 10.0. These results are possibly due to the decrease of the electrostatic and hydrophobic interactions between 1-NAA and *β*CD-PEI with the increasing pH of the medium.

1-NAA in Micellar Media

Micellar solubilization of 1-NAA in nonionic and ionic surfactants heteromicroenvironment has been monitored by fluorescence and absorption spectral techniques (Jangir and Acharya, 2014). The increase of the fluorescence intensity of 1-NAA in micellar media suggests that the suspended hydrophobic 1-NAA molecules have been solubilized.

1-NAA in β-CD Polymer (PβCD) Inclusion Complex

Dai and Kim presented the formation of the inclusion complex of 1-NAA in *β*-CD polymer (PβCD) (Dai and Kim 2012). P*β*CD was prepared by reacting *β*-CD and Epichlorohydrin (EPI) at the molar ratio of 1:7 in a strong alkalic condition. Inclusion complexes of P*β*CD and diaminobenzophenone (DABP, an alkalic aromatic compound) (DABP-P*β*CD) and 1-NAA (NAA-P*β*CD) were prepared in P*β*CD solution (1% (w/w), pH 5.0). Monoolein (MO) cubic phase containing DABP-P*β*CD and NAA-P*β*CD was prepared by a melt-hydration method. The release of FITC-dextran from the cubic phase was somewhat suppressed at neutral condition, possibly due to an extensive complexation between DABP-P*β*CD and NAA-P*β*CD in the water channels of the cubic phase.

Conclusion

The formation of inclusion compounds of auxins with CDs or other suitable molecules could improve their bioavailability, enhance their stability, reduce considerably the side effects following administration and sustain their activity for a longer time period due to their slow release. In this chapter, studies on the molecular encapsulation of eight auxins (1-NAA, 2-NAA, 4-CPA, MCPA, 2-NOA, IBA, 2,4-D and 2,4,5-T) have been extensively reviewed. The molecules that have been used as hosts are mainly native β-CD, methylated (DIMEB, TRIMEB) or hydroxypropyl β-CD derivatives and in a few cases α- and γ-CD.

The combined use of different analytical techniques has provided a complete characterization of the inclusion complexes. Their structure has been accurately determined in crystalline state by X-ray crystallography. Complex formation in solution was obtained using kneading, co-evaporation or spray-drying processing methods. NMR, spectrofluorometric and phase solubility studies verified the complexes formation in solution, determined the stoichiometry of the complexes and estimated their binding constants. The powders obtained after each processing method were studied in many cases by FTIR and DSC or other thermal analysis methods in order to check on the formation of a true solid inclusion compound and to determine its stoichiometry in the solid state.

In the crystalline state, X-ray crystallography studies on single crystals of the complexes have revealed all the relative atomic coordinates of the guest and host molecules giving the ultimate evidence of the encapsulation, the host:guest stoichiometry and a detailed description of the inclusion geometry. Single crystals suitable for X-ray diffraction were obtained only in the cases where the host molecules of the auxins inclusion compounds are β-CD, DIMEB and TRIMEB. It is observed that in the cases of auxins/native β-CD complexes, the auxins are included inside dimeric cavities that dispose a volume more than twofold of the volume of a monomeric cavity. The dimeric cavities consist of two β-CD molecules hydrogen bonded by their O3*n*-H hydroxyl groups in a head-to-head mode. This encapsulation mode is expected considering the size of the auxins molecules. In the cases of inclusion complexes with host molecules DIMEB or TRIMEB the observed host:guest stoichiometry is always 1:1 whereas in the auxin/native β-CD complexes, where the formation of β-CD dimers is always observed, the host:guest stoichiometry is 2:1 or 2:2 depending on the size, functional groups, substituents and position of substitution in the aromatic ring of the guest auxin. The guest molecule is often found highly disordered over two or three sites inside the dimeric cavity. This is a result of the weak interactions (van der Waals, hydrogen bonds, π-π, CH-π) between host-guest and guest-guest molecules.

A comparative review of the structures of the inclusion complexes can shed light on the role of the functional groups of the encapsulated auxins and their size, as well as the kind, the number and the position of their substituents, in the inclusion process. As it has been mentioned before, the CD inclusion compounds provide a simple prototype for ligand-enzyme systems, involving mainly van der Waals and hydrophobic interactions and therefore their study can provide some insight regarding the auxins binding into the pockets of the protein receptors.

The size of the indole ring of IBA does not allow an "equatorial" inclusion of IBA in the rigid hydrophobic cavity of the *β*-CD. The crystal structure of this complex confirmed an "axial" full inclusion inside the hydrophobic cavity (Figure 6) with the butyric acid substituent protruding from the primary hydroxyls rim of the host. Because of the size and the depth of immersion of the guest molecule, only one IBA molecule can be encapsulated inside the formed *β*-CD dimeric cavity. The void of the other *β*-CD hydrophobic cavity is filled by a water molecule hydrogen bonded with the nitrogen of the pyrrole ring. On the other hand, in the case of the IBA/TRIMEB complex, the much more flexible hydrophobic cavity of TRIMEB allows a partial equatorial inclusion of the indole ring (Figure 8) whereas the butyric acid substituent is located totally outside of the host.

In the cases of the complexes with 4-CPA, MCPA, 2,4-D or 2,4,5-T as guests, the size of the phenyl ring does not cause any steric effects to their inclusion inside the rigid hydrophobic *β*-CD cavity. It is the number and the position of the phenyl substituents that plays a crucial role. Two 4-CPA molecules have been found inside the formed *β*-CD dimer in the crystalline state, stacked with π-π interactions between them in an axial inclusion mode (Figure 10). The encapsulated 4-CPA molecule in the TRIMEB flexible hydrophobic cavity has been found occupying two sites: one laying at an equatorial and the other at an axial position (Figure 11a and b). The addition of a second substituent at the position 2 of the chlorophenoxy group (a methyl group in the case of MCPA or a Chlorine atom in the case of 2,4-D) causes the following results: The methyl group of MCPA does not affect too much the inclusion geometry of the complex. MCPA is found immersed in the hydrophobic β-CD cavity at about the same depth as 4-CPA and with the same axial mode. The formed complex has a 2:2 host:guest stoichiometry with the two included guest molecules stacked with π-π interactions and entrapping a water molecule inside the dimeric cavity by hydrogen bond it with their Chlorine atoms (Figure 14). The crystal structures of MCPA inclusion complexes with the more flexible DIMEB or TRIMEB have not been determined yet. On the other hand, in the case of the 2,4-D molecule where the methyl substituent is replaced by one more Chlorine atom, the inclusion compound with β-CD presents significant differences. Only one 2,4-D molecule is found encapsulated inside the formed dimericβ-CD cavity the host:guest stoichiometry being 2:1. The sole guest molecule is immersed deeper in the dimeric cavity with his chlorosubstituent at position-2 being hydrogen bonded with two water molecules entrapped at the dimeric interface (Figure 17). In the case of the 2,4-D/DIMEB complex, the guest is only partially included in the hydrophobic cavity, its dichlorophenoxy group laying totally outside the host's cavity (Figure 19). In the 2,4-D/TRIMEB complex the dichlorophenoxy group of the guest is accommodated partially inside the cavity with an orientation similar to that of the 2,4-D/β-CD and his chlorosubstituent at the position-2 along with the carboxyl group being located outside of the cavity (Figure 20). A third chlorosubstituent at the position-5 of the phenoxy group, as in the case of the 2,4,5-T molecule, causes such steric effects that do not allow the formation of crystals of inclusion complexes in hosts having rigid cavities like β-CD or DIMEB. Thus, only the crystal structure of the 2,4,5-T included inside the flexible cavity of the TRIMEB has been determined. In this crystal structure the trichlorophenoxy group of the guest is accommodated inside the cavity oriented in such a way that the chlorosubstituent at position-2 is located inside the cavity whereas the two other chlorosubstituents (at positions 4 and 5) along with the guest's carboxyl protrude outside the cavity from the broader, secondary hydroxyls rim of the host (Figure 22).

The three auxins: 2-NOA, 2-NAA and 1-NAA, that are monosubstituted naphthalene derivatives present striking differences regarding their inclusion in CDs. These differences are mainly due to the different position of the naphthalene substitution. The auxins with naphthalene substituted at position 2 (2-NOA and 2-NAA) can be encapsulated axially in the β-CD cavity (Figures 24 and 22) whereas in the case of the naphthalene substituted at position-1 (1-NAA), steric effects do not allow a complete axial inclusion of the substituted group inside the rigid β-CD. The crystallization of the 1-NAA/β-CD complex was achieved only by using a high concentration of ethanol (etOH:H_2O 1:2 v/v) in the crystal mother liquor. Thus, the determined crystal structure revealed the formation of a *β*-CD dimer where the hydrophobic cavities host ethanol molecules whereas the 1-NAA molecule is found not being encapsulated but entrapped in a sandwich mode at the interface of the *β*-CD dimer (Figure 30). The crystal structures of the inclusion complexes of these auxins with DIMEB and TRIMEB have shown that 2-NAA retains its axial inclusion geometry inside the more flexible cavity of TRIMEB (Figure 28) and 1-NAA is able to be included equatorially inside the cavities of DIMEB and TRIMEB (Figures 31 and 32) due to the flexibility of these host molecules. Moreover, Nelson et al. studied the inclusion of methyl- and ethylnaphthalene substituted at positions 1 and 2 using life-time measurements (Nelson et al., 1987). They noticed that the lifetimes of the 1-substituted naphthalenes are shorter than those of the 2-substituted naphthalenes, suggesting that *β*-CD can only include part of the 1-substituted naphthalene molecules, while the 2-substituted naphthalenes can completely enter the cavity. When more of the naphthalene ring resides in the interior of the *β*-CD, it is likely that more protection from deactivation pathways is offered. Munoz de la Pena et al. reached at the same conclusion studying the complexes of 2-NOA and 1-NAA with *β*-CD in aqueous solutions and determined their stoichiometry and stability constants by using the changes in the fluorescence intensity of the naphthalene derivatives upon inclusion (Munoz de la Pena et al., 1993). The fact that 2-NOA is completely included while 1-NAA is only partially included into the CD cavity is in agreement with the smaller variations found in the absorption and emission spectra of 1-NAA upon inclusion and with the smaller value found for the association constant. A partial inclusion gives a weaker association and smaller variations in the spectral characteristics of the guest.

The studies of the auxin/CD inclusion complexes in solution didn't always give the same results as the determination of the complex structures in the crystalline state. For example, although the crystallographic studies of the MCPA/*β*-CD clearly indicated a 2:2 host:guest stoichiometry (2 MCPA molecules been found inside the formed *β*-CD dimeric cavity) (Triantafyllopoulou et al., 2012), the host:guest stoichiometry of complexes determined by ^{1}H NMR in solution was 2:1 for the MCPA/*β*-CD complex obtained using the kneading (KN) method and 1:1 for MCPA/*β*-CD complex obtained using the co-evaporation (CE) method (Garrido et al., 2012). Moreover, the ^{1}H NMR prediction about the orientation of the MCPA molecule inside the *β*-CD cavity was opposite to that found in the determined crystal structure of the inclusion complex. Crystallography is the ultimate method to determine directly the molecular structure and thus give in every detail the inclusion geometry of the complex. However, the model of the inclusion complex obtained by crystallography is "frozen" in the crystalline state by the crystal contacts without containing any information about the dynamic behavior of the inclusion complex in solution. Molecular Dynamics simulations may give an insight about the behavior of this structure in solution. On the other hand, the ^{1}H NMR results for the MCPA/β-CD complex in solution can be interpreted on the basis of the different techniques used to prepare the CD complexes. The experimental results obtained for a CD complex can depend on the technique used since the CD complexes are held together by weak forces. Moreover, the complexation process depends heavily on the experimental conditions (pH, cosolvent, temperature, etc.). Thus, the complexes can involve species of different stoichiometries, depending on the technique used, in this case CE and KN methods (Garrido et al., 2012).

The phase solubility studies showed that in all cases β-CDs increased the water solubility of the auxins. Although this technique does not provide a clear evidence of the inclusion phenomenon (the increase of the auxins solubility is possibly due not only to its inclusion inside the CD but to other auxins-CD arrangements as well) it is widely used as a complementary method for complex characterization in solution. Thus, by the type of the obtained phase solubility diagrams a stoichiometric ratio of 1:1 was attributed to all the examined auxin/β-CD (native, Me-β-CD or HP-β-CD) complexes and a 1:2 guest:host stoichiometry was estimated in the case of the 2,4-D/α-CD complex (Pérez-Martinez et al., 1998). Stability constants were calculated from the phase solubility diagrams and indicated the following trend: Me-β-CD $\approx$ HP-β-CD > β-CD (Garrido et al., 2012; 2014).

Although the results obtained from the NMR and/or solubility studies indicate the formation of inclusion compounds, they do not preclude the possibility that the final solid products are simply physical mixtures. For this reason, the powders obtained after each processing method were studied by different techniques, in order to check on the formation of a true solid inclusion compound and to determine their stoichiometry in the solid state. For example, Pérez-Martinez et al. analyzed the final solid product of 2,4-D/α-CD with the DSC technique in order to investigate if it was inclusion complex or just a physical mixture. By analyzing the curves obtained by this method, it was revealed that 2,4-D forms an inclusion compound with α-CD only in the cases of co-precipitate and spray-dried samples verifying that the complexation process depends on the technique used and the experimental conditions. These complexes presented a 2,4-D/α-CD stoichiometry of 1:2 (Pérez-Martinez et al., 1998). In an other case Garrido et al., in addition to the DSC studies, examined the FTIR spectra of MCPA, β-CD, its physical mixture and its complex obtained by using CE and KN methods. In the physical mixture (PM), no variations on some characteristic wavenumber values were observed whereas the spectra of the inclusion complexes obtained by CE and KN methods revealed important changes that were explained as consequences of the host/guest interactions and thus indicate the formation of an inclusion complex (Garrido et al., 2012).

In general, the studies of the complexes obtained showed different structural and thermal properties and distinct spectroscopic features from the free molecules. The gathered results allow concluding that CDs can enhance the water solubility of all the examined auxins, a fact that can be regarded as an important step in the design of novel auxin formulations with enhanced chemical properties. Moreover, these results can be regarded as an essential step to the development of controlled release agricultural formulations containing auxins.

REFERENCES

Allerup S, "Changes in Transpiration Induced by Ethyl Alcohol", *Nature* 194, no. 4834 (June 23, 1962): 1193–94, doi:10.1038/1941193b0.

Añibarro M, Gessler K, Usón I, Sheldrick GM, Harata K, Uekama K, Hirayama F, Abe Y, Saenger W,"Effect of Peracylation of B-Cyclodextrin on the Molecular Structure and on the Formation of Inclusion Complexes: An X-Ray Study", *Journal of the American Chemical Society* 123, no. 48 (November 1, 2001): 11854–62, doi:10.1021/ja010696b.

Banks JL, Beard HS, Cao Y, Cho AE, Damm W, Farid R, Felts AK, Halgren TA, Mainz DT, Maple JR, Murphy R, Philipp DM, Repasky MP, Zhang LY, Berne BJ, Friesner RA, Gallicchio E, Levy RM. "Integrated Modeling Program, Applied Chemical Theory (IMPACT)," *Journal of Computational Chemistry* 26, no. 16 (December 1, 2005): 1752–80, doi:10.1002/jcc.20292.

Bethanis K, Tzamalis P, Hountas A, Mishnev AF, Tsoucaris G, "Upgrading the Twin Variables Algorithm for Large Structures", *Acta Crystallographica Section A Foundations of Crystallography* 56, no. 2 (March 1, 2000): 105–11, doi:10.1107/S0108767399013355.

Bethanis K, Tzamalis P,Hountas A, Tsoucaris G, Kokkinou A and Mentzafos D, "New Developments of the TWIN Algorithm for Phase Extension and Refinement in Disordered Supramolecular Structures", Acta Crystallographica. Section A, Foundations of Crystallography 56, no. Pt 6 (2000): 606–8, doi:10.1107/S0108767300011314.

Brutti C, Apóstolo N. M, Ferrarotti S. A, Llorente B. E, Krymkiewicz N, "Micropropagation of *Cynara Scolymus L.* Employing Cyclodextrins to Promote Rhizogenesis", *Scientia Horticulturae* 83, no. 1 (2000): 1–10, doi:10.1016/S0304-4238(99)00067-9.

Beurskens P. T, Beurskens G, de Gelder R, García-Granda S, Gould R. O. & Smits J. M. M. Th*e* DIRDIF 2008 *Program System* (2008). Crystallography Laboratory, University of Nijmegen, The Netherlands.

Cai Y, Tarr M. A, Xu G,Yalcin T, Cole R. B, "Dication induced stabilization of gas-phase ternary beta-cyclodextrin inclusion complexes observed by electrospray mass spectrometry", *Journal of the American Society for Mass Spectrometry*, 14(5), 449-459, (2003).

Cavallaro V, Trotta F, Gennari M, Di Silvestro I, Pellegrino A, Barbera A. C, "Effects of the complex Nanosponges-Naphthaleneacetic acid and *β*-Cyclodextrins on in vitro rhizogeneseis of globe artichoke" *Acta Horticulturae* (ISHS) 983:369-372 (2013).

Challa R, Ahuja A, Ali J, Khar R. K,"Cyclodextrins in drug delivery: an updated review", *AAPS Pharm Sci Tech*. Jun 2005; 6(2): E329–E357, doi: 10.1208/pt060243.

Connors K. A, "Binding Constants, the Measurement of Molecular Complex Stability", New York: John Wiley & Sons, (1987).

Cremer, D., and Pople J., A, "General Definition of Ring Puckering Coordinates", *Journal of the American Chemical Society* 97, no. 6 (March 1, 1975): 1354–58, doi:10.1021/ja00839a011.

Dai J, Kim J. C, "Monoolein Cubic Phase Containing Ionizable Aromatic Compounds-Loaded β-Cyclodextrin Polymers: FITC-Dextran Release Property*", Advanced Science Letters*, Volume 5, Number 1, January 2012, pp. 247-252(6) Publisher: American Scientific Publishers (2012).

Del Valle E. M. M, "Cyclodextrins and Their Uses: A Review," *Process Biochemistry*, no. 9 (2004): 1033–46, doi:10.1016/S0032-9592(03)00258-9.

Dodziuk E, "Modeling of CDs and Their Complexes" ch.11, pp. 333, in "Cyclodextrins and Their Complexes. Chemistry, Analytical Methods, Applications", Weinheim, Ed. H. H. Dodziuk, Wiley-VCH, 2006.

Ejchart A. and Kozminski W, "NMR of Cyclodextrins and Their Complexes" ch.9, pp. 231, in "Cyclodextrins and Their Complexes. Chemistry, Analytical Methods, Applications", Weinheim, Ed. H. H. Dodziuk, Wiley-VCH, 2006.

Ermondi G, Cillis G. I, Caron G,"In silico strategies to describe the formation of the inclusion complex between b-cyclodextrin and b-naphtyloxyacetic acid: a preliminary step towards prediction of log *K*" *J. Incl. Phenom. Macrocycl. Chem.* 57:355–361 (2007).

Farrugia L. J, "*WinGX* Suite for Small-Molecule Single-Crystal Crystallography", *Journal of Applied Crystallography* 32, no. 4 (August 1, 1999): 837–38, doi:10.1107/S0021889899006020.

Fourtaka A, "Inclusion of bioactive compounds in cyclodextrins", MSc thesis, Department of Science, Agricultural University of Athens, (2012).

Guilfoyle T, Hagen G, Ulmasov T, Murfett J, "How Does Auxin Turn On Genes?", *Plant Physiology* 118, no. 2 (October 1, 1998): 341–47, doi:10.1104/pp.118.2.341.

Kabsch W, "Automatic Processing of Rotation Diffraction Data from Crystals of Initially Unknown Symmetry and Cell Constants", *Journal of Applied Crystallography* 26, no. 6 (December 1, 1993): 795–800, doi:10.1107/S0021889893005588.

Garrido E. M, Marcos Santos, Silva P, Cagide F, Garrido J. and Borges F, "Host-Guest Complexes of Phenoxy Alkyl Acid Herbicides and Cyclodextrins. MCPA and B-Cyclodextrin", *Journal of Environmental Science and Health, Part B* 47, no. 9 (2012): 869–75, doi:10.1080/03601234.2012.693867.

Garrido J, Cagide F, Melle-Franco M, Borges F. and Garrido E. M, "Microencapsulation of Herbicide MCPA with Native B-Cyclodextrin and Its Methyl and Hydroxypropyl Derivatives: An Experimental and Theoretical Investigation", *Journal of Molecular Structure* 1061 (March 5, 2014): 76–81, doi:10.1016/j.molstruc.2013.12.067.

Giordano F, Novak C and Moyano J. R, "Thermal Analysis of Cyclodextrins and Their Inclusion Compounds", *Thermochimica Acta, New Advances in Pharmaceutical Thermal Analysis*, 380, no. 2 (December 14, 2001): 123–51, doi:10.1016/S0040-6031(01)00665-7.

Guo P, Su Y, Cheng Q, Pan Q, Li H. "Crystal Structure Determination of the B-Cyclodextrin-P-Aminobenzoic Acid Inclusion Complex from Powder X-Ray Diffraction Data," *Carbohydrate Research* 346, no. 7 (May 15, 2011): 986–90, doi:10.1016/j.carres.2011.03.003.

Hedges A. R, "Industrial application of cyclodextrins", *Chem. Rev.* (June 25 1998), 98(5), pp 2035–2044, doi:10.1021/cr970014w.

Higuchi T, Connors K. A, "Phase solubility techniques", *Adv. Anal. Chem. Instrum.* 1965, 4, 117 – 212.4.

Hiller E, Krascsenits Z, Čerňanský S, "Sorption of Acetochlor, Atrazine, 2,4-D, Chlorotoluron, MCPA, and Trifluralin in Six Soils From Slovakia", *Bulletin of Environmental Contamination and Toxicology* 80, no. 5 (May 1, 2008): 412–16, doi:10.1007/s00128-008-9430-9.

Hou T, Wang J, Li Y, Wang W, "Assessing the Performance of the MM/PBSA and MM/GBSA Methods. 1. The Accuracy of Binding Free Energy Calculations Based on Molecular Dynamics Simulations", *Journal of Chemical Information and Modeling* 51, no. 1 (November 30, 2010): 69–82, doi:10.1021/ci100275a.

Jangir S. K. and Acharya S, "Studies on 1-Naphthaleneacetic acid in micellar media by spectrofuometry", *International Journal of Basic and Applied Chemical Sciences,* Vol. 4 (1) January-March, pp.18-22 (2014).

Jorgensen W. L, Maxwell D. S, Tirado-Rives J, "Development and Testing of the OPLS All-Atom Force Field on Conformational Energetics and Properties of Organic Liquids", *Journal of the American Chemical Society* 118, no. 45 (January 1, 1996): 11225–36, doi:10.1021/ja9621760.

Lee M. S. and Kim J. C, "Microgels formed by electrostatic and hydrophobic interaction of naphthaleneacetic acid with β-cyclodextrin-grafted polyethyleneimine", *Colloid Polym. Sci.* (2011) 289:1177-1183 doi: 10.1007/s00396-011-2442-8.

Liu J. and Li S. F. Y, "Separation and Determination of Auxins by Capillary Electrophoresis", *Journal of Liquid Chromatography & amp; Related Technologies* 19, no. 11 (1996): 1697–1713, doi:10.1080/10826079608013998.

Mazi V, "Formation, crystal structure determination and bioassays of the indole-3-butyric acid inclusion compound in β-cyclodextrin", MSc thesis, Department of Science, Agricultural University of Athens, (2012).

Mark P. and Nilsson L, "Structure and Dynamics of the TIP3P, SPC, and SPC/E Water Models at 298 K", *The Journal of Physical Chemistry A* 105, no. 43 (October 6, 2001): 9954–60, doi:10.1021/jp003020w.

Mentzafos D, Mavridis I. M, Le Bas G, Tsoucaris G,"Structure of the 4-Tert-Butylbenzyl Alcohol–β-Cyclodextrin Complex. Common Features in the Geometry of B-Cyclodextrin Dimeric Complexes", *Acta Crystallographica Section B Structural Science* 47, no. 5 (October 1, 1991): 746–57, doi:10.1107/S010876819100366X.

Mura P, Ceccarelli L, Faucci M. T, Rinaldelli E, Mancuso S,"Improvement of solubility Of indolbutyric acid by complexation with alpha-cyclodextrin and rhizogenic activity in *Olea europea* L. cv. Leccio del Corno", *Advance Horticultural Science* 9, (1995):119:121.

Nelson G, Patonay G, Warner I. M, "Fluorescence Lifetime Study of Cyclodextrin Complexes of Substituted Naphthalenes" *Applied Spectroscopy Volume 41, Number 7 (Oct. 1987)* pp. 1235-1238.

Nuzzo A, Scherman O. A, Mazzei P. and Piccolo A,"pH-Controlled Release of Auxin Plant Hormones from cucurbit[7]uril Macrocycle", *Chemical and Biological Technologies in Agriculture* 1, no. 1 (December 1, 2014): 1–8, doi:10.1186/2196-5641-1-2.

Otwinowski Z. and Minor W, "Processing of X-Ray Diffraction Data Collected in Oscillation Mode", *Methods in Enzymology*, ed. Jr. Charles W. Carter, vol. Volume 276, Macromolecular Crystallography Part A (Academic Press, 1997), 307–26.doi:10.1016/S0076-6879(97)76066-X.

Pérez-Martínez J. I, Arias M. J, Ginés J. M, Moyano J. R, Morillo E, Sánchez-Soto P. J. and Novák Cs, "2,4-D-alpha-cyclodextrin complexes. Preparation and characterization by thermal analysis", *Journal of Thermal Analysis* 51(3): 965-972 (1998).

Ponder J. W. and Richards F. M, "An Efficient Newton-like Method for Molecular Mechanics Energy Minimization of Large Molecules", *Journal of Computational Chemistry* 8, no. 7 (October 1, 1987): 1016–24, doi:10.1002/jcc.540080710.

Qiu D, Shenkin P. S, Hollinger F. P. and Still W. C, "The GB/SA Continuum Model for Solvation. A Fast Analytical Method for the Calculation of Approximate Born Radii", *The Journal of Physical Chemistry A* 101, no. 16 (April 1, 1997): 3005–14, doi:10.1021/jp961992r.

Quigley D. and Probert M. I. J, "Constant Pressure Langevin Dynamics: Theory and Application", *Computer Physics Communications* 169, no. 1–3 (July 2005): 322–25, doi:10.1016/j.cpc.2005.03.072.

Rekharsky M. V. and Inoue Y, "Microcalorimetry", ch.8, pp.199, in Dodziuk H, "Cyclodextrins and Their Complexes: Chemistry, Analytical Methods, Applications", Weinheim, Ed. H. H. Dodziuk , Wiley- VCH, 2006.

Ryckaert J. P, Ciccotti G, and Berendsen H. J. C, "Numerical Integration of the Cartesian Equations of Motion of a System with Constraints: Molecular Dynamics of N-Alkanes", *Journal of Computational Physics* 23, no. 3 (March 1977): 327–41, doi:10.1016/0021-9991(77)90098-5.

Roberts H. A, "Weed Control Handbook: Principles", no. Ed. 7: 533pp. Oxford, Blackwell Scientific Publications, 1982.

Sheldrick G, "A Short History of *SHELX*", *Acta Crystallographica Section A Foundations of Crystallography* 64, no. 1 (January 1, 2008): 112–22, doi:10.1107/S0108767307043930.

Thireou T, Atlamazoglou V, Levakis M, Eliopoulos E, Hountas A, Tsoucaris G and Bethanis K, "CrystTwiV: A Webserver for Automated Phase Extension and Refinement in X-Ray Crystallography", *Nucleic Acids Research* 35, no. suppl 2 (July 1, 2007): W718–W722, doi:10.1093/nar/gkm225.

Triantafyllopoulou V, Mentzafos D, Tsorteki F, Bethanis K, "Inclusion Compounds of Plant Growth Regulators in Cyclodextrins. Part VI. Study of the Inclusion Compound of MCPA in B-Cyclodextrin by X-Ray Crystallography", *Journal of Chemical Crystallography* 42, no. 3 (March 1, 2012): 238–44, doi:10.1007/s10870-011-0231-3.

Triantafyllopoulou V, Tsorteki F, Mentzafos D. and Bethanis K, "Inclusion Compounds of Plant Growth Regulators in Cyclodextrins, Part VII: Study of the Crystal Structures of 2-Naphthylacetic Acid Encapsulated in B-Cyclodextrin and heptakis(2,3,6-Tri-O-Methyl)-B-Cyclodextrin Complexes by X-Ray Crystallography", *Journal of Inclusion Phenomena and Macrocyclic Chemistry* 75, no. 3–4 (April 1, 2013): 303–10, doi:10.1007/s10847-012-0164-2.

Triantafyllopoulou V, "Crystal structure analysis of inclusion compounds", PhD thesis, Department of Biotechnology, Agricultural University of Athens, (2013).

Tsorteki F. and Mentzafos D, "Structure of the Complex of heptakis(2,6-Di-*O*-Methyl)-β-Cyclodextrin with (2,4-Dichlorophenoxy)acetic Acid", *Carbohydrate Research* 337, no. 13 (2002): 1224 – 1228.

Tsorteki F, Bethanis K. and Mentzafos D, "Structure of the Inclusion Complexes of heptakis(2,3,6-Tri-O-Methyl)-Beta-Cyclodextrin with Indole-3-Butyric Acid and 2,4-Dichlorophenoxyacetic Acid", *Carbohydrate Research* 339, no. 2 (January 22, 2004): 233–40.

Tsorteki F, Bethanis K, Pinotsis N, Giastas P, Mentzafos D, "Inclusion Compounds of Plant Growth Regulators in Cyclodextrins. V. 4-Chlorophenoxyacetic Acid Encapsulated in Beta-Cyclodextrin and heptakis(2,3,6-Tri-O-Methyl)-Beta-Cyclodextrin", *Acta Crystallographica. Section B, Structural Science* 61, no. Pt 2 (April 2005): 207–17, doi:10.1107/S0108768105004106.

Uekama K, Hirayama F, Irie T. "Cyclodextrin Drug Carrier Systems",*Chemical Reviews* 98, no. 5 (July 30, 1998): 2045–76.doi:10.1021/cr970025p.

Van Uden, Woerdenbag H. J. and Pras N, "Cyclodextrins as a Useful Tool for Bioconversions in Plant Cell Biotechnology", *Plant Cell, Tissue and Organ Culture* 38, no. 2–3 (September 1, 1994): 103–13, doi:10.1007/BF00033867.

Wang E, Chen G, Han C, "Crystal Structure of a Novel Sandwich Inclusion Complex of β-Cyclodextrin with a-Naphthylacetic Acid" *Chemical Research in Chinese Universities* 27, no. 5 (2011): 730-733.

Yang X, Kim J. C, "β-Cyclodextrin Hydrogels Containing Naphthaleneacetic Acid for pH-Sensitive Release", *Biotechnol Bioeng*. Jun 1;106(2):295-302. (2010) doi: 10.1002/bit.22696. PMID:20148415.

Yannakopoulou K, Kokkinou A, "Structure of the Complex of Beta-Cyclodextrin with Beta-Naphthyloxyacetic Acid in the Solid State and in Aqueous Solution", *Carbohydrate Research* 332, no. 1 (2001): 85–94, doi:10.1016/S0008-6215(01)00072-6.

Yurkova M. P, Pozdnyakov I. P, Plyusnin V. F, Grivin V. P, "Photochemistry of Herbicide 2,4,5-Trichlorophenoxyacetic Acid in Aqueous Solutions in the Presence of Cyclodextrins" Chemistry for Sustainable Development 19, 659-664 (2011).

Yurkova M. P, Pozdnyakov I. P, Plyusnin V. F, Grivin V. P, Bazhin N. M, Kruppa A. I. and Maksimova T. A, "A mechanistic study of the photodegradation of herbicide 2,4,5-trichlorophenoxyacetic acid in aqueous solution", *Photochem. Photobiol. Sci.*, 12(4), 684-689, (2013).

INDEX

A

B

C

D

E

F

G

H

I

N

O

P

Q

R

S

T

U

V

W

X

Y